Systems Thinking and Systems Engineering
Volume 1

A Journey Through the
Systems Landscape

Volume 1
A Journey Through the Systems Landscape
Harold "Bud" Lawson

Systems Thinking and Systems Engineering Series Editors
Harold "Bud" Lawson   bud@lawson.se
Jon P. Wade   jon.wade@stevens.edu

A Journey Through the Systems Landscape

Harold "Bud" Lawson

ISBN 978-1-84890-010-3

College Publications
Scientific Director: Dov Gabbay
Managing Director: Jane Spurr
Department of Computer Science
King's College London, Strand, London WC2R 2LS, UK

http://www.collegepublications.co.uk

Printed by Lightning Source, Milton Keynes, UK

Contents

Interlude 2: Case Study in Organizational Development...............89

Chapter 3 Acting in Terms of Systems 99

Chapter 7 Data, Information and Knowledge 219

Interlude 4: Case Study in Ontology Life Cycle Management 237

Chapter 8 Organizations and Enterprises as Systems...251

Preface

The ability to "think" and "act" in terms of systems is a prerequisite to being able to lead and operate private and public organizations and their enterprises so that their purpose, goals and missions can be effectively and efficiently pursued. Thinking in terms of systems is intimately bound with the ability to understand the structure of systems as well as the behavioral interrelationships of multiple systems in operational situations. Systems thinking, also called the systemic approach, has evolved through a variety of contributions beginning in the 1920s into a discipline that can be applied in gaining an understanding of the common denominator aspects of various types of systems and, in particular, the dynamic relationships between multiple systems in operation.

Through systems thinking organizations and their enterprises can learn to identify system problems and opportunities and to determine the need for, as well as to evaluate the potential effect of, system changes. Having decided upon the need for new systems, removal of systems and/or structural changes in one or more existing systems, it is vital to deploy a controlled means of "acting" for managing the changes in an expedient and reliable manner. In this regard, the principles of systems engineering including well defined system life cycle management processes such as those defined in the international standard ISO/IEC 15288 (System Life Cycle Processes) provides relevant guidance for the life cycle management of any type of man-made system.

This book has been developed to convey essential properties of organizational system assets, understanding the treatment of dynamic system situations as well as to focusing upon one of the most central activities of any organization or enterprise; namely the management of change. This is accomplished by introducing the concepts and principles of systems from the perspective of systems (systemic) thinking as well as from a system engineering perspective and the guidance provided by the ISO/IEC 15288 standard. A model for change management based upon paradigms for thinking and acting as well as for the gathering of knowledge is provided. The model can be applied anywhere in an organization or enterprise where there is a need for critical decision making in respect to treating problems and/or pursuing opportunities related to system assets and/or system situations.

A primary goal of this book is to provide insight for individuals and to promote communication within groups and teams that have a vested interest in understanding complex system situations and responses as well as improving upon the management of the system assets utilized in achieving their organization or enterprise purpose, goals, and missions. The ultimate goal is to assist in creating a learning organization that continually improves its capabilities to think and act in terms of systems.

THE JOURNEY

The book is organized as a journey through the conceptual systems landscape of an organization and its enterprises. The journey is made through highly modular chapters in which important concepts and principles are presented as the knowledge required for thinking and acting in terms of systems. In order to promote knowledge assimilation, each chapter terminates with a number of questions and exercises that are provided to assist the reader in verifying their knowledge of the chapter contents. This approach can and should be used for group or team learning situations as well since the questions and exercises definitely will lead to useful discussions, dialogue and collective learning. The figure on the following page outlines the journey.

Introduction to Systems – The journey begins identifying the omnipresence of systems, the steps that have been taken in the systems movement as well as the interdisciplinary nature of systems. A unifying model that indicates the central role of structure and behavior in scientific, engineering and other disciplines is presented and the disciplines of Systems Thinking and Systems Engineering are introduced. Systems are then categorized into fundamental types. Hierarchies and networks are identified as the two rudimentary system topologies. Multiple viewpoints and views of systems are presented. The perspective that systems are not real and only exist as descriptions is presented as a controversial argument. The achievement of organization and enterprise purpose, goals, and missions based upon the usage of systems as assets is described. System assets are described as sustained systems and are contrasted with situation systems arising from problems or opportunities as well as respondent systems created to handle opportunity or problem situations.

**The journey about
to be taken**

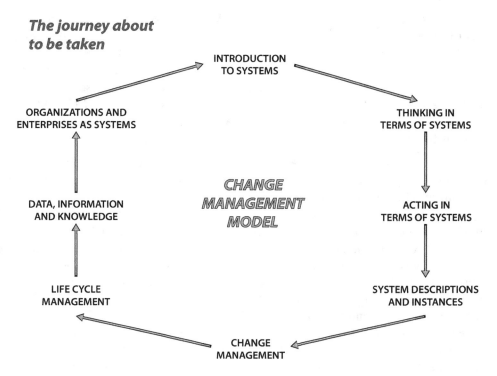

The collection of individual operative systems into a system-of-systems to meet a complex crises situation or to form an extended enterprise is described. A generic change model is introduced which focuses upon both structural and operational changes as well as the gathering of knowledge with both feedback and feed forward mechanisms. Various sources of system complexities and the need for a holistic view of systems are presented. Finally, the informal introduction to systems provided in this first chapter is formalized into a Systems Survival Kit composed of a small number of concrete concepts, a universal mental model and a small number of principles. The reader should always keep this kit in mind since it can and will be used to describe the treatment of all types of systems.

Thinking in Terms of Systems –Systems thinking which has been evolving since the 1920s into a discipline that is also referred to as systemic thinking is introduced. The primary focus of systems thinking is upon utilizing a holistic perspective in understanding the dynamics of interaction amongst multiple systems during their operation. It is via this perspective that underlying problems and opportunities can be identified. The goals of systems thinking are presented along with some of the "tools" of the trade. The difference between hard systems and soft systems is described. Modeling is a central aspect of systems thinking for which there are a variety of methods and tools ranging from structured text and various graphic representations used for qualitative analysis to models decomposed to programming languages and used as a basis for quantitative analysis via simulation. Several approaches are described including the Five Why's method, Influence Diagrams, Peter Senge's Links, Loops and Delay language, Rich Pictures, Systemigrams, STELLA and iThink. Peter Checkland, one of the foreground figures in the field of systems thinking provides a link between thinking and acting in what he calls the Soft System Methodology (SSM). The major thrust of Action Research which leads to the need for SSM as well as the SSM model introduced by Checkland are considered. Finally, a few general principles to keep in mind in relationship to systems thinking are presented.

Acting in Terms of Systems – A paradigm that utilizes two well-known loops; namely OODA (Observe, Orient, Decide, Act) and PDCA (Plan, Do, Check, Act) is introduced. These loops are integrated into the change management model to deal with the continuous activities associated with situation awareness and decision-making; respectively the discrete project related activities associated with making changes in a controlled and reliable manner. The background of and importance of the systems engineering discipline for accomplishing changes in systems is presented. The life cycle management of systems via the usage of processes is central to systems engineering. The organization of system-of-interest life cycles in respect to stages and supporting enabling systems is described. The ISO/IEC 15288 standard that has been developed in order to provide a basis for world trade in system products and services is introduced. The standard provides a means of defining systems, defining the bounds of systems as well as identifying key processes involved in the life cycle management of systems. An example of the usage of the standard is provided as well as a description of how the standard is to be tailored in order to meet the specific needs of organizations, enterprises, projects and agreements. By providing clear system relevant definitions and addressing the needs of all system actors, the standard promotes systems thinking; however, its major contribution is the vital guidance it provides for acting in terms of systems.

System Descriptions and Instances – This chapter emphasizes the importance of understanding the fundamental differences between descriptions of systems and system instances in the form of products and services. First a perspective on viewing the important work products of processes as successive versions of the System-of-Interest is introduced. The importance of the establishment of driving concept and principles as well as achieving balance in the use of architecture, processes, methods and tools during the life cycle is emphasized. Next, life cycles are reduced to reflect three fundamental transformations (Definition, Production, and Utilization) and the universal mental model is applied in establishing situation goals and objectives for building stage related projects. Various important aspects of the life cycle are described including the project scope, transforming requirements to architecture, baselines and configurations, produced products and operational aspects are described. The importance of system architecture is highlighted by considering the main features of the International Standard ISO/IEC 42010 (Architecture Definition). The concepts and principles of a Light-Weight

Architectural Framework (LAF) that provides for describing the architectural related work products of important system actors is introduced. Next, the important question of the ownership of system descriptions (definitions), products and services is presented and the affect of trading in system products and services in portrayed. Finally supply chain relationships that are established in the trading of system products and services within various life cycle stages are presented.

Change Management – The ability to organize and operate change activities is the topic of this chapter. The feedback model of a cybernetic system composed of a controlling element, a controlled element, and a measurement element is explained. The utilization of cybernetics in an organization as introduced by Stafford Beer is presented. It is then shown that the Change Management Model is, in fact, a cybernetic system. The measurement of effect of a change is essential in order to determine if threshold goals have been attained and in determining the need for further changes. Various types of measurements for products, services and processes are reviewed. The importance of consistent decision-making is described as well as the consequences of improper decision-making, the so-called entropy effect. Guidance is provided on how change management can be implemented within the framework of the ISO/IEC 15288 standard. Finally a refinement of the OODA part of the Change Model, in particular a variant called DOODA (Dynamic OODA) aimed at rapid decision-making in Command and Control operations is described.

Life Cycle Management of Systems – This chapter begins by considering the differences between Management and Leadership in respect to systems. The role of Change Control Boards in respect to the universal mental model is described. Various types of systems having varying life cycle lengths are presented and illustrated. The topic of System-of-Systems is reexamined and ownership issues are discussed. A deeper look at life cycle models based upon several portrayals in the form of a T-model provides insight into how to apply the ISO/IEC 15288 standard in various types of situations. The concepts of iterative and development and implementation of systems as well as progressive and incremental acquisition are described. The properties of the well known Vee Model and Spiral Model are illustrated. The roles and responsibilities to be taken by various actors during a life cycle are presented. An approach to integrating life cycle models with processes along with methods and tools is presented. Finally, a description of product life cycles from an enterprise perspective is provided and the relationship between system and product life cycle management, including Integrated Logistics Systems, is clarified.

Data, Information and Knowledge – It is vital to assimilate knowledge related to systems that can be used in making wise decisions about change. Thus, understanding the significance of as well as the relationship between data, information, knowledge and even wisdom is addressed in this chapter. The view of information also includes various forms of computerized multi-media. The quality of information that has become a vital issue for both thinking and acting in terms of systems is considered and the classification of information according to taxonomies and ontologies is described. The gathering of data, information and knowledge during system life cycles is described as a vital contributor to the intellectual capital of an organization and its enterprises. The importance of building information models is presented. The role of creative thinking as a means of stimulating new knowledge related to problems and opportunities is described and illustrated. Finally, the five disciplines described by Peter Senge that are the keystones of a learning organization; namely personal mastery, mental models, shared vision, team learning, and systems thinking are reviewed.

Organizations and Enterprises as Systems – At this last stop on the journey it becomes clear that an organization and/or its enterprises are systems composed of system elements and relationships and thus must be life cycle managed as well. The role of managers as system owners is described. The view of the architecture of an enterprise as an aggregate of organization related system architectures are presented. To deal with the growing complexity of enterprise architectures, the application of the Light-Weight Architecture Framework (LAF) is proposed. A strategy for leading change in an organization is presented including a discussion of why system changes often fail to achieve their goals. An indication of how the journey through the systems landscape has contributed to the achievement of ISO 9000 quality management principles; namely, customer focus, leadership, involvement of people, process approach, system approach to management, continual improvement, factual approach to decision making, and mutually beneficial supplier relationships is presented. Finally, a strategy for implementing management system standards such as ISO 9001 and ISO 14001 is presented.

In summing it all up, conclusions are drawn in respect to the benefits accrued due to the sharing of a common discipline independent view of systems in the systems landscape of an organization and its enterprises.

Interludes – In order to gain a deeper understanding of how to apply the concepts and principles of "thinking" and "acting" in terms of systems, a number of case studies are provided at various points in the book. Three of the case studies are based upon project work performed by course participants and one case study is based upon architectural work performed by your author. The interludes are as follows:

Case Study in Crises Management

Case Study in Organizational Development

Case Study in Architectural Concepts and Principles

Case Study in Life Cycle Managing Ontologies

ACKNOWLEDGMENTS

A number of people have contributed to this book both directly and indirectly. Since life is a learning experience, the author is fortunate to have been exposed to a wide range of system-related problems and opportunities during his professional career that began in the late 1950s. Thus, I would like to thank all of my colleagues with which I have worked in the computer industry, at academic institutions, as well as in my systems related consulting activities at many public and private organizations for their direct and indirect contributions to my own systems knowledge. In particular, I am pleased to acknowledge the strong influence of my first boss and mentor; namely the legendary late Rear Admiral Dr. Grace Murray Hopper. Grace, a pioneer of the computer industry, taught me to be exploratory, to seek deeper insight and to always question the status quo which proved to be an excellent start for my own journey into the world of complex systems.

In 1996 I became involved as head of the Swedish delegation and in 1999 also as the elected architect of the ISO/IEC 15288 (System Life Cycle Processes) standard developed within ISO/IEC JTC1 SC7 Working Group 7. My participation in Working Group 7 was at the request of Dr. Raghu Singh and I gratefully acknowledge his insight in initiating this important international standards project. Participation in this effort was sponsored by the Swedish Defence Materiel Administration (FMV) and by the Swedish development agencies NUTEK and VINNOVA. In this regard, I acknowledge the continual support of Ingemar Carlsson and others at FMV and the support of Karl-Einar Sjödin of NUTEK and later VIN-NOVA. Within the Working Group 7 project, I wish to thank all of my colleagues for the many hours of fruitful discussions at meetings in all corners of the world. In particular, I thank Stuart Arnold, editor of the standard for our close cooperation in establishing and "defending" the concepts and principles upon which the architecture of 15288 rests as well as Stan Magee and Doug Thiele who managed the effort in an expedient manner.

There have been several people that have made important contributions to systems (systemic) thinking. A web search on this topic yields thousands of relevant references. In particular, I have been inspired by the contributions of Peter Senge in his pioneering work on "The Fifth Discipline"; namely Systems Thinking and the disciplines required to establish a learning organization. Also, I have drawn upon the work of Robert Flood who has provided deeper insight into work of Senge and others, including his own contributions in his book "Rethinking the Fifth Discipline." My colleague at Stevens Institute of Technology, John Boardman has made important contributions to developing diagrams representing complex situations in for form of Systemigrams that are presented in the thought provoking book "Systems Thinking: Coping with 21st Century Problems" co-authored by Brian Sauser. Peter Checklands contributions in recognizing how to apply systems thinking to non-technical systems, that is, the Soft Systems Methodolgy (SSM) which evolved over many years has been an inspiration. The contributions of other systems thinkers including Russel Ackoff, Ross Ashby, Staffard Beer, Jay Forrester and Wes Churchman have also influenced the presentation in the book. More recently, I have been inspired by the work of another pioneer in the systems thinking discipline; namely, Georgy Petrovich Schedrovitsky who was early to present this topic in Russia.

In recent years, I have had the opportunity to explore the central issues of change management. In this regard, I wish to acknowledge the close cooperation of Johan Bendz of the Swedish Defence Materiel Administration in formulating early versions of a change management model as well as for our many deep discussions on critical system issues. Together with Lennart Castenhag of Svenska Kraftnät and Gösta Enberg of the Stockholm County Government some important ideas of applying ISO/IEC 15288 to IT Management were developed in the Egiden project. My thanks to Dr. Dinesh Verma of the Stevens Institute of Technology for suggesting that I develop a graduate Systems Thinking graduate course at Stevens Institute of Technology where early versions of the book have been utilized. Also, I wish to thank Jack Robinson and his colleagues at the IT department of the Stockholm County Government for their interest in applying the concepts provided in early versions of the book.

The graduate and professional development courses that led to this book have been delivered several times in the USA and Sweden and I am grateful to Sten and Anita Andler at Skövde University, Anita Kollerbauer at Stockholms University, Andreas Ermedahl at Mälardalen University, Peter Gabrielsson at the Swedish Defence Materiel Administration, Berndt Brehmer, Per-Arne Persson and Mats Persson at the Swedish National Military College as well as Jan-Inge Svensson at the Folke Bernadotte Academy for their support in organizing and delivering the course in Sweden.

To all the course participants, I express my deep appreciation for teaching me a lot about fields in which I had no prior knowledge. As reflected in the many projects that have been done, systems are truly universal. Some of the project results are included at various points in the book and in particular as three case studies. I thank my Syntell AB colleagues Stuart Allison, Jonas Andersson, Ulf Carlsson, Mike Cost, Asmus Pandikow, Tom Strandberg and others for many interesting discussions. Thanks to the Syntell AB managing director Mats Bjorkeroth for providing a consulting environment in which the concepts and ideas presented in this book are applied in a variety of concrete organizational and enterprise situations. Finally, many thanks to Mats Persson for his vital assistance in preparing figures and formatting of the book as well as to my Stevens Institute of Technology colleague Jon Wade for his thorough review of the book.

Special acknowledgement is given to the late Christer Jäderlund, a Swedish computer pioneer and a true professional in thinking and acting in terms of systems. He often likened systems to a kaleidoscope as portrayed on the cover of this book. That is, that depending on how you turn it, the system view as seen by the beholder can have different structures and yield different behavioral experiences. As the reader pursues the material in this book, this kaleidoscope perspective of systems will successively become evident.

Finally, I thank my wife Annika and children Adrian and Jasmine for their patience, support and assistance during the many years of providing courses on the subject matter that has led to the production of this book.

Happy reading and understanding during the journey.

Harold "Bud" Lawson

Lidingö, Sweden

Chapter 1
An Introduction to Systems

"There's so much talk about the system and so little understanding."
Pirsig, R.M., Zen and the Art of Motorcycle Maintenance, 1974

There are very few words that can be interpreted in so many ways as the word "system". What is a system? This is most often a question of perspective. Yet, we all use this term to describe something that is essential. The solar system, the weather system, the energy system, the political system, the school system, the hardware system, the software system, the automotive system, the financial system, the sanitary system, the management system, the urban planning system, the legal system, the social system, and so on. It is quite clear that systems, although often abstract in nature, are in some sense present and affect us continually. It is important to note that some of the systems, like the solar system and the weather system are the works of nature whereas the other exemplified systems are all man-made.

Our understanding of systems, particularly complex ones is at best cursory as noted in the quotation from Pirsig [Pirsig, 1974]. For all but trivial systems complete understanding is virtually impossible. So, we live with the fact that our understanding lies somewhere between mystery and mastery [Flood, 1998]. This uncertainty often causes an uneasy feeling about systems. Via the journey taken in this book, the reader will be able to remove much of the mystery and take a positive step towards systems awareness and at least partial mastery.

SYSTEMS ARE EVERYWHERE

Ludwig von Bertalanffy [von Bertalanffy, 1968], an Austrian biologist considered by many as the father of modern systems thinking, points to the fact that systems are everywhere. We may not always formalize our view of systems, but we certainly feel the effect of them. None of us will ever forget the system related influences in the international financial crises during the fall of 2008. Systems that are intimately related can have strong causal influences upon each other. Let us first consider the background to the identification and formalization of systems, the most fundamental concepts of systems as well as omnipresence of systems in all fields of endeavor.

THE SYSTEMS MOVEMENT

During the 20th century, a number of key contributions were made to the systems area. In particular, during and after World War II, there was a growing awareness of the importance to examine and understand complex entities composed of multiple elements in a holistic manner. This movement is still gaining momentum and is attracting the attention of prominent researchers and practitioners. Given the complexities of modern society one can ask the questions: Why has it taken so long time to achieve focus upon this vital area? Is there an active systems movement? How is it being realized?

Focusing upon the holistic properties of entities is not a new phenomenon. In fact, the Greek philosophers, in particular Aristotle pointed to this need in examining multiple factors in explaining the universe. Thus his works in physics, logic, metaphysics, ethics, politics, and biology included observations on the need to treat holistic properties. This very early holistic view survived up to the 17th century. Then came the scientific revolution. Driven by the need to prove or disprove a specific hypothesis, scientific method began to evolve in the works of amongst others Kepler, Galileo, Bacon and Descartes.

The scientific methods developed from the 17th century and onward can be characterized by the need to isolate one or a small number of elements of the phenomenon that is to be studied. This view of reducing to elements that can be studied in isolation and a hypothesis that can be proved or disproved has in fact hindered the development of holistic systems thinking. There were, of course, some exceptions where a broader view of natural phenomenon was considered and that have lead to a broader understanding of natural laws. It was Isaac Newton that provided the first scientific explanation of the universe in terms of motions of the earth and moon that led to Newton's invention of calculus as a mathematical tool. The Newtonian view prevailed up to the major paradigm shift in science introduced by important generalizations provided by Albert Einstein in his theory of relativity.

It was during the 1920s that Ludwig von Bertalanffy pointed to analogies between holistic properties of biological and other systems and the current systems movement got its start. Von Bertalanffy went on to apply his observations to a wide variety of systems including management systems and organizations [von

Bertalanffy, 1968]. Checkland [Checkland, 1993] as well as Skyttner [Skyttner, 2001] provide excellent historical summaries of systems thinking and as well as the scientific movement starting from the early Greek contributions that have led up to the contemporary views of systems.

It is now clear that there is an active systems movement even though it is difficult to gain a precise agreement on what it is, what is included, and even what, if anything, should be excluded. In this book we shall examine some of the central developments of the systems movement to see how it is being expressed in theory and applied in practice.

FUNDAMENTAL PROPERTIES

In this chapter, we introduce a set of concepts and principles that will enable your ability to "think" and "act" in terms of systems. The understanding and usage of the concepts and principles is considered to be the most vital aspect of the book as it will affect your own ability to see the system aspects of any type of system as well as to communicate with others concerning system related problems and opportunities. We begin with the most fundamental concept.

> *"We believe that the essence of a system is <u>togetherness</u>, the drawing together of various parts and the relationships they form in order to produce a new whole ..."*
> **John Boardman and Brian Sauser** [Boardman and Sauser, 2008]

This first fundamental concept of "togetherness" permits us to recognize as von Bertalanffy postulated that systems are everywhere. The notion of togetherness leads us to two further concepts; namely <u>structure</u> and <u>behavior</u>.

Structures and behaviors are central properties of all man-made systems. The structure of a system is a static property and refers to the constituent elements of the system and their relationship to each other. The behavior is a dynamic property and refers to the effect produced by a system "in operation."

Another related fundamental property that is attributed to systems is the concept of emergence. Emergence arises from both the predictable and unpredictable operational behavior of a system itself and/or in relationship to the environment in which the system resides. This concept is captured in the following quote by Peter Checkland.

> *"Whole entities exhibit properties which are meaningful only when attributed to the whole, not to its parts . . ."*
> **Peter Checkland** [Checkland, 1999]

DISCIPLINE INDEPENDENCE

The omnipresence of systems implies that understanding system properties and utilizing them is independent of the discipline in which systems are considered. For complex systems, it is the collective understanding of the dynamics of system behavior as well as the life cycle management aspects of systems that is often, of necessity, the result of interdisciplinary efforts. In order to neutralize the discipline effect and focus upon "system content", it is vital to unify *thinking* and *acting* on the part of individuals and groups coming from diverse specialist backgrounds and possessing diverse knowledge, skills and capabilities. In this regard, important unifying aspects are portrayed in Figure 1-1.

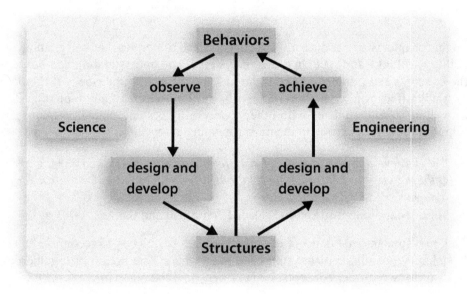

Figure 1-1: Science and Engineering Relationship to Structures and Behaviors

Both scientific and engineering related disciplines deal with the fundamental system concepts of structures and behaviors. In the case of scientific disciplines, the scientist observes behaviors (in nature or in man-made systems) and then attempts to find and describe structures (in some form of "language") that explain the behaviors. In the case of engineering disciplines, the engineer based upon the need to provide required (specified) behaviors, designs and develops structures that when produced and instantiated achieve behavioral requirements.

To illustrate the difference of approach to structures and behaviors, consider the following disciplines, some of which are traditionally associated with natural sciences, others that have used science in the name of the discipline and a wide-variety of engineering disciplines:

(x) Science	(y) Engineering
Biological	Electrical
Physical	Mechanical
Chemical	Chemical
Environmental	Sanitation
Management	Business Process
Computer	Software
System	Systems
Health	Health Care
Military	Military

As an exercise, the reader can consider how these disciplines map into the scientific and engineering view of structures and behaviors portrayed in Figure 1-1. While these discipline examples have a scientific or engineering relationship to structures and behaviors, it may not be as obvious in other disciplines. For example, it is interesting to speculate about how *art* is related to science and engineering. There are at least two possible relationships:

- Aesthetically pleasing structures that are appreciated in the "eyes of the beholder". For example, in nature, a rainbow is a pleasing structure. To a mathematician, the structure of a proof may be pleasing. For a software engineer, a clear algorithm that provides a desired behavior in a non-complex manner may be pleasing.

- Another relationship comes from the term "artisan". The term artisan is typically applied for someone who is mature in his/her discipline. Most typically, artisans are able to design and develop structures that meet needs and thus are most similar to the engineering profession. However, true artisans are most always capable of observing then finding and describing relevant structures.

The artistic relationship introduces the important notion of style into systems related work. The reader is encouraged to consider other relationships between art and science as well as art and engineering. Further, consider structural and behavioral relationships in disciplines such as medicine, psychology, sociology or other disciplines with which you are familiar.

SYSTEM THINKING AND SYSTEMS ENGINEERING

It becomes clear that the work of all disciplines can be related to some aspect of systems. In fact, we are all systems thinkers and system engineers in the sense that we are constantly thinking and acting in responding to the system situations that affect our every day lives. Understanding the core concepts of the disciplines of System Thinking and System Engineering in theory and in practice provides the means for making systems a focal point (first class object) that can be exploited in improving our ability to deal with complex systems in any field of endeavor.

Thinking in terms of systems is highly related to observing the dynamic behaviors of systems in operation and thus correlates to the left-hand (scientific) side of Figure 1-1. However, as opposed to the scientific method involving reduction of behaviors into elements to be studied in isolation systems thinking is based upon observing and describing the holistic behaviors of multiple systems and their system elements.

Acting in terms of systems involves creating (engineering) structures of one or more systems that are of interest and thus is highly related to the right-hand (engineering) side of Figure 1-1. This leads us naturally into the focus of this journey through the systems landscape; namely the coupling of systems thinking and systems engineering. Actually, they are quite related. Simply utilizing systems thinking without learning to evaluate alternative structural improvements and establish goals and plans for improvements in systems does not make sense. On the other side, acting in terms of systems via system engineering without understanding the underlying reasons for and implications of acting does not make sense. So, the natural coupling of thinking and acting in terms of systems leads to the need for decision making and change management that will be considered in depth as we proceed through the journey of the systems landscape.

CLASSIFYING SYSTEMS

A taxonomy would be a useful tool in structuring the system journey to be taken in this book. Such a complete enumeration of systems is in general not possible since the perspective on systems is highly context dependent. On the other hand, for practical purposes the enumeration of systems that are of interest for a particular purpose is quite important and achievable. In lieu of a comprehensive taxonomy and to focus upon various types of systems, the classification by Checkland [Checkland, 1993] provides a useful starting point. The reader will observe that systems can be placed into one or more of these four categories.

Natural systems – These systems have their origin is in the universe and are as they are as a result of forces and processes which characterize the universe. They are systems that could not be other than they are, given a universe whose patterns and laws are not erratic.

Defined physical systems – These systems are the result of conscious design aimed at satisfying some human purpose. They are composed of physical elements that have well defined relationships.

Defined abstract systems – These systems do not contain any physical artifacts but are designed by humans to serve some explanatory purpose. Abstract systems can include mathematical descriptions, poems or philosophies. They represent the ordered conscious product of the human mind. Definitions of systems composed of function and/or capability elements are examples of abstractions that can later be captured in other man-made system forms, physical or as concrete human activities.

Human activity systems – These systems are observable in the world of innumerable sets of human activities that are more or less consciously ordered in wholes as a result of some underlying purpose or mission. At one extreme is a system consisting of a human wielding a hammer, at the other international political systems that are needed if life is to remain tolerable on our small planet. This will include a priori defined sets of processes composed of activities (not explicitly addressed by Checkland) as well as sets of activities viewed from a particular perspective of interested parties.

Note that software systems are a hybrid between defined abstract and defined physical systems in that from abstract descriptions in some form of language or model, program code is generated by a translator program that when integrated with and executed by a computer (defined physical system) creates an emergent behavior. The term software-intensive system is also used to describe a system that consist primarily of software but also contains other physical and often human activity elements.

In this book, focus is placed upon the man-made systems and system situations that are of importance for individuals as well as for various groups including public and private organizations and their enterprises in developing capabilities for learning to think and act in terms of systems. Thus, understanding defined physical, defined abstract, human activity and software systems are all important for achieving this goal. Natural systems are, of course, not excluded since natural occurring elements may be incorporated as elements of a man-made system or as elements in the environment in which a man-made systems operates.

SYSTEM TOPOLOGIES

There are two fundamental topologies for systems that form the basis of "togetherness"; namely the hierarchy and the network as illustrated in Figure 1-2.

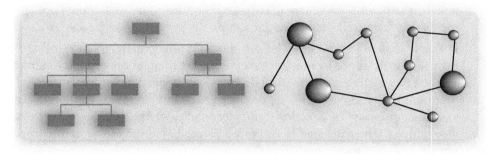

Figure 1-2: Hierarchy and Network System Topologies

The hierarchy topology is the result of a defined system that is developed to meet some need. The system results from an analysis that decomposes a system into constituent elements at two or more levels. This decomposition leads to a logical basis for understanding, partitioning, developing, packaging, and managing the system in a prudent manner. This topology is typical for the planned development of products (physical and/or abstract), but also can be found in the planned development of an organization, enterprise or even project. Such human activity organization chart usage of hierarchy is quite common for explaining who has responsibility for parts of the system, work to be performed on the system, as well as for establishing a chain of command (who reports to who).

The network topology can be used to capture essential properties of defined physical systems; for example networks of plumbing, highways, train tracks, power transmission, telecommunications and, of course, the internet. At a higher level, network topologies can capture defined abstractions such as capabilities or functions to be provided and as stated earlier can then form the basis for physical system realizations. Such systems physical or abstract are typically designed for change; that is, the topology is changed over time where nodes and/or links are added or removed.

The network topology is also relevant for human activity systems including social systems where various forms of relationships between human elements (individuals and/or groups) can be expressed. Such systems may or may not be planned. If they are planned they can be used to regulate relationships. However, they can arise due to elements and relationships that evolve and in this case attempt to portray, even difficult, conflicting interpersonal relationships. Networks arise due to a problematic situation when multiple elements interact in a manner that is dangerous. For example, a terrorist, a bomb, a subway, and passengers become the elements of a dangerous network of elements and relationships.

The two system topologies are not exclusive in and of themselves. It is quite clear that an organization described as a hierarchy does not always function according to a strict line of command. Networks, even though undefined, arise between individuals and groups that provide the necessary elements and relationships to get things done. Further it is clear that individual elements in a physical network such as a transformer in a power grid are products that deliver services and have been planned and developed as systems for their individual purpose or need. These elements are systems, in their own right, that can be decomposed, developed and managed according to a hierarchy.

MULTIPLE VIEWPOINTS AND VIEWS

"A system is a way of looking at the world…a system, any system, is the point of view of one or several observers."

Gerald Weinberg [Weinberg, 2001]

In line with the togetherness property expressed by Boardman and Sauser as well as the perspective noted by Weinberg, any collective set of elements (parts) that have some form of relationship can be viewed as a system. Based upon perspectives and vested interests and concerns of stakeholders (i.e. viewpoints), individuals, groups, teams, organizations and enterprises, will view systems in different manners. This kaleidoscope aspect of systems is exemplified in Figure 1-3.

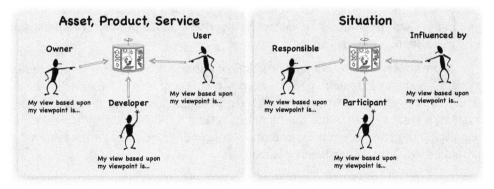

Figure 1-3: Multiple Viewpoints and Views

Defined physical, defined abstract, software and even certain human activity systems may be viewed by some as assets, be viewed by others as products, and by still others as value added services. Associated with these views and based upon their roles and responsibilities as individuals, groups, teams, organizations and enterprises, they have viewpoints about a defined system reflecting their vested interest concerns (for example, as owner, acquirer, developer, user or maintainer). Such systems are planned, developed and utilized in order to achieve some defined purpose.

In contrast to planned systems, situation systems arise due to dynamic interactions of multiple systems in operation (including natural systems). Such systems can be based upon human activity such as a man wielding a hammer, a political situation, an emergency situation or crises that has arisen (for example fire, a tsunami, hurricane, terror action, and so on). As noted in the figure, once again there can be different views reflecting viewpoints that are based upon concerns related to the system situation (for example, as responsible for the occurrence of the situation or responsible for responding to the situation, participating in the situation, or by being influenced by the situation).

DO SYSTEMS REALLY EXIST?

Regardless of the viewpoints and views related to systems or their topology, one can ask an interesting question:

Do systems really exist?

This may seem to be a philosophical issue, but let us use this perspective to illustrate a viewpoint about systems. In the Checkland categorization provided earlier, it was observed that natural systems are as they are (i.e. they exist). However, all other forms of planned or situation systems; namely, defined physical, defined abstract and human activity systems are either the creative planned products of humans or due to a situation that has arisen.

More concretely, the planned products of humans, an aircraft system, a motor system, an airframe system, and so on, may conjure up visions of something that really exists and can be "touched". On the other hand, a political system, a school system, a legal system, an urban planning system, while abstractly representing something of great importance, cannot be "touched". Thus, what is the system? A potentially *debatable* perspective is that:

Man-made systems exist only as descriptions.

Your author has often utilized the elements displayed in Figure 1-4 as a basis for focusing upon this debatable perspective. You are now challenged. Are these elements a system? Why or why not?

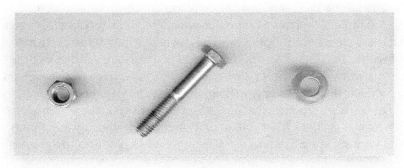

Figure 1-4: Bolt, Nut and Lock Washer

After thinking about this question, consider a secondary question. That is, to determine if these elements as they lie have any purpose? Do they fulfill any need?

Further think of the individual or groups that have responsibility for designing and for producing each of the individual elements illustrated in the figure. Do they view them as systems? Do they view them as products? Or do they view them as the services that they potentially deliver? Or do they view them as all three?

Consider then physically putting the elements together with two or more additional objects (containing appropriate holes) in order to fasten these objects. Has a system been constructed? You may think so, but consider the fact that in order to create this construction, the elements (including the objects to be fastened) have been defined and these instances of the objects, the nuts, the bolts and the lock washers have been produced according to some description (specifications). There can also be a description of the assembly procedure to put the physical elements together. Given the definition of the elements, their relationships and the assembly procedure, multiple instances of the assembly can be achieved. Have we not produced products based upon the system description? Thus if we want to use the terms product and system to denote two different concepts, we must accept that the system is really a description and thus the "system" does not exist. Let us examine this line of reasoning.

Planned man-made hierarchical or network systems are composed of defined elements and defined interrelationships. At best hardware, software or human elements of a system can be viewed as real objects that can be in some sense "touched". However, element existence in a planned system is based solely upon their description as hardware elements, software elements, human elements and element interrelationships.

In the case of a value added product the system description serves as the "template" from which product instances can be generated (singular production at one extreme up to being mass-produced at the other extreme). In an analogous manner, value added services, for example, a banking service, are the result instantiating service operation following the system description of the service as a "template".

To further illustrate this point of view concerning systems consider the following concrete example. The laptop computer upon which this book was prepared is a product; the system description of which is owned by and is life cycle managed by a manufacturer that integrates its elements. The hardware elements of the system may be owned by other parties that manage the life cycles of these elements as system products that they supply to computer system integrators. Further, there are a variety of software products that operate on the hardware, the system descriptions of which as well as their life cycle management are owned by supplying organizations. These software systems are also supplied as products to the integrator.

Thus, the defined representation of assets, products and services varies at various points during the life cycle. At early stages in the life cycle the described system is typically viewed as a defined abstract system composed of a set of functions and/or capabilities that have defined relationships. As the design is realized into a deliverable product and/or service, the system description becomes more specific, either in the form of physical elements and in the form of defined activities for humans (procedures or processes) or combinations of both.

In respect to situations that arise, it is only when we decide to think about or concretely to describe the elements of a situation and their interrelationships that system-like properties become apparent. Otherwise it is just a situation. For complex situations such descriptions if attempted are seldom complete and once again are based upon views of the situation reflecting the viewpoints and concerns of parties involved in the situation.

In summary, one can take a view that systems only exist as descriptions. However, as noted in Figure 1-3 and the discussion above, a defined system can be viewed as an asset by some people viewed as a product by others, and still others as the service it provides. So, it boils down to a question of concerns and viewpoints as to whether a product is indeed a system or a service is a system or they are simply products and services. Or alternatively, to avoid confusion it may be useful to differentiate between systems as descriptions, system products and system services. We will not belabor this philosophical point other than to indicate that once again our viewpoints and views can effect how a system is observed.

SYSTEMS-OF-INTEREST

All forms of man-made systems and natural systems as well potentially contain large numbers of elements as pointed to the following:

> *"At this point, we must be clear about how a system is to be defined. Our first impulse is to point at the pendulum and to say "the system is that thing there." This method, however, has a fundamental disadvantage: every material object contains no less than an infinity of variables, and therefore of possible systems. The real pendulum, for instance, has not only length and position; it has also mass, temperature, electric conductivity, crystalline structure, chemical impurities, some radioactivity, velocity, reflecting power, tensile strength, a surface film of moisture, bacterial contamination, an optical absorption, elasticity, shape, specific gravity, and so on and on. Any suggestion that we should study all the facts is unrealistic, and actually the attempt is never made. What is necessary is that we should pick out and study the facts that are relevant to some main interest that is already given."*
> **W.R. Ashby** [Ashby, 1956]

Thus it is important to identify: Where is your system-of-interest? What are its salient elements? How is it related to other systems and to the environment in which it is contained? These are vital questions to ask. Flood and Carson [1998] provide a useful perspective in this respect as portrayed in Figure 1-5.

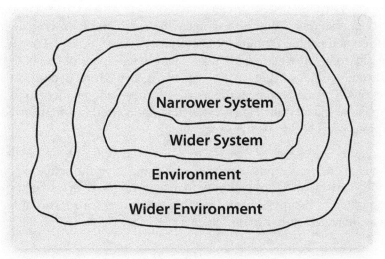

Figure 1-5: Systems-of-Interest in their Environment(s)

A system can be categorized as being a *closed system* in which no elements of the system are found to have relationships with anything external to it. For example, a perpetual motion machine that continues to operate based upon counterbalancing weights without any influence from the environment in which it operates. In contrast, an *open system* is characterized by exchanges of material, information and/or energy between itself and its environment across a boundary.

Thus for open systems while we might focus upon the elements and relationships of a Narrower System-of-Interest (NSOI) we must also consider its context in terms of a Wider System-of-Interest (WSOI) and well as the environment(s) in which they operate. Let us consider two examples:

A business enterprise that sells toys is a system composed of planning, marketing and sales, management, research and development, production and distribution elements. Thus the business can be considered as a Narrow System-of-Interest (NSOI) upon which we can focus. However, it is part of a Wider System-of-Interest (WSOI) that encompasses, amongst other elements, their customers as well as their suppliers of raw materials. The business is operated in an environment where the toys are marketed and changes in that environment due to consumer attitudes towards the toys, economic factors, competitors and so on will have an affect upon the Wider System-of-Interest and in turn the narrower toy business System-of-Interest. There is also a wider environment that can also affect the closer environment as well as the other Systems-of-Interest. For example, toy safety regulations that can affect the consumption of the toys.

As another example of the relationships portrayed in Figure 1-5, consider an action composed of a terrorist, a bomb, a subway, and passengers as elements and relationships of this dangerous situation. This narrower (NSOI) is tied to a wider (WSOI) by amongst other elements, contacts with a terrorist organization, the supply of materials, know how to make the bomb, the subway system and the composition and mental framework of the passengers. The NSOI and WSOI exist in an environment where there is a system based upon for example political, economic and religious beliefs as well as intelligence efforts to discover potential terror actions. This environment in turn is encompassed in wider environment in which decisions in the form of laws and regulations concerning political, economic, and religious aspects are taken into account.

The reader will observe that via these two examples as well as the earlier discussions about viewpoints, concerns and views, that the scope of systems is quite wide. This broad scope certainly indicates that there is a vested interest in removing much of the mystery and moving towards at least a partial mastery of systems as indicated earlier in this chapter.

SYSTEM ASSETS

There are a variety of planned man-made systems in operation within all types of organizations (public, private and even non-profit). These planned systems are essential to endeavors (enterprises) that work towards achieving purpose, goals and missions as portrayed in Figure 1-6. Thus, the enterprise as well as the organization must focus upon *(institutionalize)* its portfolio of system assets. The availability of, as well as the condition (that is, state) of these assets is an essential aspect of managing the organization and its enterprises. Some of the portfolio assets are the value added system products and/or services that the enterprise produces; other system assets are those utilized in supporting the enterprise in its operations by providing essential infrastructure services. The ISO/IEC 15288 standard was developed in order to provide guidance to all types of organizations and their enterprises in managing the life cycles of man-made systems resulting in products or services or as enabling infrastructure systems [ISO/IEC, 2002 and 2008]. Thus, the standard can be applied for the management of defined physical, defined abstract as well as human activity systems.

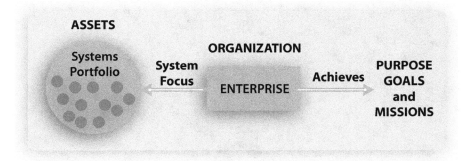

Figure 1-6: Achieving Organization Purpose, Goals and Missions via System Assets

In this book, the terms organization and enterprise are used interchangeably [Note 1-1]. It is clear that an enterprise always has an organization and that an organization is also an enterprise. Further, it is clear that as large complex organizational conglomerates (extended enterprises) have arisen, in the private and public sectors, the quantity and scope of system assets as well as the integration of and institutionalizing of these assets has led to many system related complexities. In order to avoid the continual repetition of these two important terms, the term enterprise is most often utilized to be a single enterprise up to and including a large extended enterprise. The reader can, except when explicitly indicated, for all practical purposes view the terms organization and enterprise as being synonymous.

Needs, Services and Effect

The man made systems that an enterprise provides as value added products and/ or services as well as the systems they utilize as infrastructure assets have been designed to meet some form of need as portrayed in Figure 1-7. In this figure and as described Figure 1-5, the system upon which focus is placed is as noted earlier the *System-of-Interest* (SOI). The SOI is designed to provide services to the user and when an instance of the system product or system service is deployed, it delivers effect.

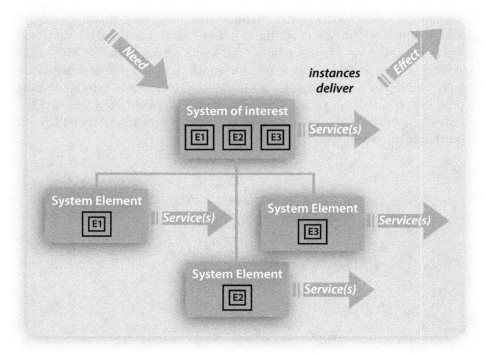

Figure 1-7: System Structure: Need, Services and Effect

For example, my laptop computer (as a product) has been designed to meet the needs of multiple users. It offers a wide variety of services to users, but in writing this book the computer and I are operating, that is interacting together, as elements of a human activity SOI. This SOI is planned in order to deliver a desired effect; namely producing this book as a product.

The *system elements*, in the figure; namely, E1, E2, or E3, are integrated in expectation that they will contribute to meeting the need to be provided by the SOI. Each of the system elements can deliver one or more services to the SOI and when interacting with the other elements of a system product or service instance an effect emerges.

Implicit in Figure 1-7, are the central system related notions of *structure* and *behavior*. That is, the SOI portrayed has a structure that is defined by the set of system elements it contains as well as the relationships defined between the elements. The services that can be provided are the potential emergent behavior of the system. When the system product and/or service are used in operation to meet a need it produces effect that is the actual behavior of the system.

It is important to note that the effect delivered (i.e. behavior) for the system product or service is not simply the individual behaviors of the system elements. The behavior that results from the operation of the interacting elements of the system is thus, as introduced earlier, called the *emergent behavior*. The reader should compare these system properties with the earlier discussion of the nut, bolt, lock washer and objects to be fastened together.

My laptop computer is composed of hardware and software elements each of which provide services and deliver effect. I, as another element in this SOI working in cooperation with the computer element, am able to produce a behavior that leads to the production of this book. None of these elements alone could have achieved that behavior. Thus, the behavior is indeed, emergent.

Humans in general can have a variety of relationships to planned systems. They can be stakeholders in a system and thus expect that a SOI provide services and deliver effect that is in their interest. Humans can operate an instance of an SOI and thus be an element in that system. Finally, they can be part of an operational environment in which they interact with one or more systems; that is, they are consumers of services provided by systems, supply services to other systems or are simply influenced by the system.

The SOI portrayed in Figure 1-7 enumerates the static structure of system elements that are part of a hierarchical system topology. This *structural view* is only one view of the system. To be more explicit about the system elements and their dynamic behavioral relationships, a network topology is required to portray an *operational view* as illustrated in Figure 1-8.

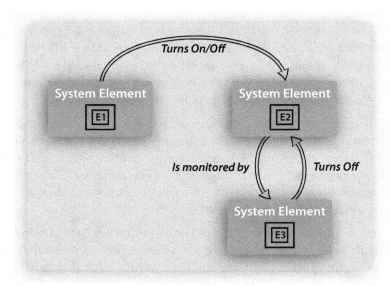

Figure 1-8: System Elements and Behavioral Relationships

In this description we see explicit relationships where the relationships that are offered as services are identified. We could use this example of a behavioral model (operational view) for many concrete physical system instances. For example where E1 is a climate-control switch, E2 is a heating or cooling element, or both, and E3 is a thermostat that monitors the temperature. These fundamental relationships are provided in a variety of products such as space heaters, toasters and a variety of other household appliances. This network representation identifies the potential behaviors of the system in contrast to the static hierarchical system element enumeration illustrated in Figure 1-7.

To illustrate another SOI consider E1 being an operator setting up and then turning on a duplicating machine, E2 and E3 being a software controlled element in the duplicating machine that turns off the machine when finished or when some form of serious malfunction occurs. The reader can certainly relate this type of "control" structure to a variety of familiar system products.

As noted earlier, at various points in the life cycle of a system the description varies. At earlier stages, the elements may be defined as functions or capabilities with relationships. At later stages, these definitions are refined to concrete integrated elements of hardware, software or human activities that provide the function and/or capability.

Decomposition

An SOI defined abstractly or concretely typically includes system elements that are themselves systems and therefore contain their own system elements as portrayed in Figure 1-9. These lower level systems can be further decomposed into system elements that are themselves, in turn, systems. The decomposition of systems in this hierarchical manner is called *recursive* decomposition and is a central concept of the ISO/IEC 15288 standard. At each level, one or more of the system elements may themselves be systems. The standard handles this recursive decomposition in a very consistent manner. At each level portrayed in Figure 1-9 where system elements are systems of a lower level, the standard is reapplied at that level to provide for the integration of system elements as the SOI of that level. Thus, the SOI is level dependent and changes when different levels are considered for life cycle management.

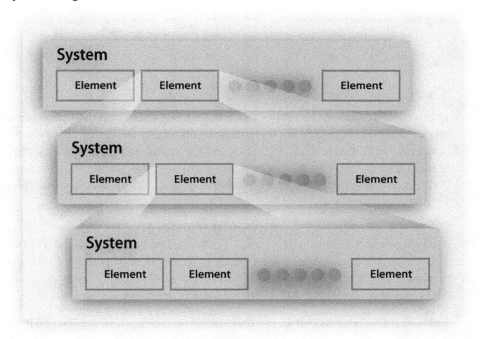

Figure 1-9: System Structure: Level-wise Decomposition

In the recursive decomposition portrayed in Figure 1-9, there are three levels; consequently, there are three levels each containing one or more Systems-of-Interest. The ISO/IEC 15288 standard is reapplied at each of the three levels, perhaps by another supplier enterprise in the life cycle management of systems-of-interest of the level.

At some point, the decomposition of systems into system elements terminates. Thus, there is a *stopping rule* that is related to practical need as well as the risks associated with the system element. That is to say, if there is no advantage to further decomposition and/or the system element is well defined and can be incorporated with controlled risks (perhaps can be purchased as a standard "off-the-shelf" element or its provision is guaranteed without further decomposition), the recursive decomposition can be terminated.

Let us return to my laptop computer. The SOI that is important to me as the author is composed of the computer system and myself. In this context, I have no practical need to consider further decomposition of these two elements. On the other hand, the computer system is a System-of-Interest of a supplier that integrates its elements. In turn the hardware system and the software system elements are Systems-of-Interest products that are life cycle managed by others. Both the hardware and software systems are further decomposed into system elements that are systems, and so on. Thus the respective owners of these Systems-of-Interest apply the stopping rule according their practical needs and the risks involved in provisioning of the elements.

Typical System Assets

It is vital for public, private and even non-profit organizations and their enterprises to understand and agree upon what their institutionalized system portfolio assets are and how they are related to each other. For this purpose, a categorization is useful. While the set of specific assets varies amongst private and public organizations and their enterprises, the categories of defined systems assets portrayed in Table 1-1 are usually discernable.

All public and private enterprises exist in order to provide some form of value added product(s) and/or service(s). These systems along with the systems for product/service management are typically the main (narrow) NSOI focus of an enterprise. However, in the wider WSOI context, all of the other systems are "enablers" for the provisioning of system products and/or system services.

Even though not necessarily viewed by enterprise personnel in an explicit manner each one of these institutionalized assets forms a system composed of system elements and relationships that must, in some manner, be life cycle managed. Formally life cycle managing these systems according to appropriate life cycle models makes them explicit and leads to an improved understanding of the nature of systems and their role in the enterprise [ISO/IEC, 2002 and 2008]. That is, those responsible for an asset as well as those influenced by the system asset develop a shared view of systems with other enterprise asset responsible parties. Such clear understanding and assignment of system asset responsibilities is a prerequisite to the effective operation of private, public and even non-profit enterprises.

Table 1-1: Institutionalized Organization/Enterprise System Assets

A. Value-Added Product Related	– Personnel
B. Value-Added Service Related	– Supply Chain
C. Product/Service Management	– Agreements
– Development	– Contracts
– Production	– Life Cycle Models
– Operation	**E. Organization Related**
– Support	– The Enterprise
– Disposal	– Division
D. Enterprise Operation	– Department
– Policy	– Project
– Management	– Task Force
– Financial	**F. Information Related**
– Marketing	– Data and Information
– Facilities	– Information Processing
– Process	– Knowledge
	– Knowledge Management

System Elements

The system elements of a planned System of Interest asset may be of various types including those illustrated in Table 1-2.

Table 1-2: Potential System Element Types

Hardware – mechanical, electronic	**Procedures** – operating instructions
Software – system software, firmware, application, utilities	**Facilities** – containers, buildings, instruments, tools
Humans – activities, operators	**Natural elements** – water, gas, organisms, minerals, and so on
Data – individual items and sets of data	**Other types of elements** – policies, laws, regulations, patents, contracts, agreements
Information – data items or sets of data that have defined interpretations	
Processes – business, political, system management	

Most of the system elements that are enumerated are familiar as elements in man-made systems; however the last two categories deserve further explanation.

As mentioned earlier, naturally occurring elements such as water, gas, air, organisms, minerals, and so on can be included as elements in a man-made system. For example an automotive system while composed of a large number of hardware and software elements and a human element, also requires the usage of water, gas and air as system elements in order to operate.

This list of system elements includes other element types that are of importance (concern) to particular parties from their viewpoint. For example, policies, laws, regulations, patents, contracts, and agreements may be considered as environmental factors affecting systems to some parties, but to others, they are considered as elements of an SOI. They may even be considered as systems; for example an agreement can be viewed as a SOI to those involved with acquisition of system products or system services.

SUSTAINED, SITUATION/RESPONDENT AND THEMATIC SYSTEMS

Depending upon the type of value added product or service that the public, private or non-profit enterprise supplies their provisioning related and enabling system assets have varying longevity. Institutionalized systems must be properly *sustained* over long periods of time in order to be in such condition that when put into operation (instantiated) are ready to deliver the desired effect.

The provisioning of value added products and services such as aircraft, telecommunication equipment, banking services, health care, social welfare, etc. requires a long sustained life cycle. Typically such sustained systems result in product or service families. So from a generic system description, variant products and services are produced, each one of which must be life cycle managed.

Systems can arise as a *situation* that may be short-term but may have a long longevity. The situation may be thought of and even described in terms of a network of contributing elements and relationships as illustrated in the terrorist action described earlier. In order to counter-act the situation that has arisen a *respondent* system is created and put into operation. For example, consider as a respondent system, a fire brigade that is assembled from elements (equipment, consumables (water, chemicals, etc.), and personnel) in order to bring a fire under control. Another example of a respondent system is the assembly of a military force in order to pursue a Course of Action to meet a situation that has arisen. Such system services are composed from available assets (equipment and people) and form a temporary system asset that is defined quickly and put into operation by a mission related task force. During the operation of the system service feedback concerning situation developments are used to rapidly restructure (redefine and dimension) the respondent system in order to meet changing needs.

Situation systems also arise in the operation of any organization and represent a challenge to the organization in putting together a respondent system. Typically the situation is met by formation of a task force or project that will handle the situation, be it a problem (perhaps crises), or an opportunity for the organization. Depending upon views and viewpoints, the situation and respondent system can be seen as coupled into a single larger WSOI where the systems (situation and respondent) interact.

In relating situation systems to respondent systems and sustained system assets, consider the introduction of a system-coupling diagram as portrayed in Figure 1-10.

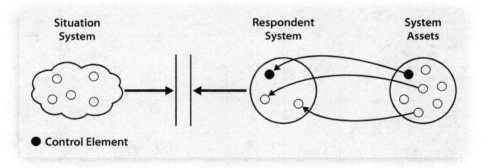

Figure 1-10: System-Coupling Diagram

Here we see clearly the formation of a respondent system based upon system assets. One of the elements to be incorporated must be a control element that directs the respondent system in its operational activities in responding to the situation system. The situation system provides both input to the respondent system and is the recipient of outputs from respondent system actions. The reader should keep this coupling diagram in mind, as we will return to it several times during the journey. It should be a familiar scenario for everybody. Consider the situation of getting somewhere by some means of public or private transport. We are always building respondent systems in our mind based upon system assets such a knowledge of routes, available transportation media, time schedules, and so on. Indeed as von Bertalanffy stated, systems are everywhere.

The situations described above are *real*, that is they actually occur. Another form of situation system is *thematic*. That is, they are constructed for the purpose of studying the systemic aspects of a potential problem or opportunity situation as a theme. That is (what if ?) a particular problem situation or opportunity arises. In addition to studying the problem or opportunity situation one or more respondent systems may also be created in order to study the effect produced by potential courses of action or to actually practice in the form of a simulated situation/response environment. Such training sessions are quite common in military environments and for civilian crises management. They can also be used as a basis for business games and management exercises in any type of organization.

The nature of the real situation or thematic situation can be related to some aspect of the structure or behavior of the system assets of an organization, or be related to network interrelationships that have evolved as a system. Further, in putting together a respondent system, instances of system assets are deployed as elements. The coupling of a set of elements from diverse system assets for several situations is indicated in Table 1-3. Implicit in the coupling is the creation of a network topology.

Table 1-3: Elements of Real or Thematic Situations and/or Respondent Systems

Situation Systems \ Institutionalized Systems	Value Added Product Related	Value Added Service Related	Product/ Service Management	Enterprise Operation	Enterprise Related	Information Related	Environment
Problem with Declining Sales	•	•	•	• •	• •	•	• •
Opportunity for a New Product Line	• •	•	• •	• •	• •	• •	•
Problem with Labor Union Relationship			• •	• •	• •	• •	•
Problem with Management Conflict		• •	• •	• •	• • •	• • •	• •

The reader will observe the treatment of typical organization related problems and opportunities. Of particular importance is the Environment column. It is often due to the events in the environment (narrow and/or wider) in which instantiated systems operate where problems and/or opportunities arise.

The treatment of problem and opportunity situations in the context of international operations established for the purpose of creating peace and stability in countries in which some form of turmoil situation exists forms a vital system. In this context, elements from multiple spheres are referred to by the acronym PMESII (Political, Military, Economic, Social, Infrastructure and Information) [Joint Publication 2.0, 2007]. The coupling of elements between the spheres is illustrated in Figure 1-11. The network can represent the system coupling of elements contributing to or being affected by a problem situation, or the elements of a respondent system to meet the problem, or both.

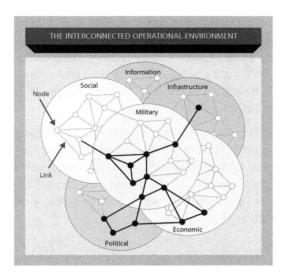

Figure 1-11: Network of Political, Military, Economic, Social, Infrastructure and Information Elements

The composition of real situation or thematic situation systems from the elements of multiple systems and their interrelationships is often temporary and thus the respondent systems are typically not defined as sustained assets and are typically not life cycle managed. However, some respondent systems treating longer term problems or opportunities may have a long life time in which case they should also be life cycle managed.

Thematic systems are used as the object of study and learning in order to determine the need for change in the institutionalized systems. The system principles described earlier in respect to system elements and relationships are the same for sustained, real situation and thematic situation as well as respondent systems. All of these forms of systems are of interest to public, private and non-profit enterprises and should be treated holistically by systems thinking methodologies as described in Chapter 2.

SYSTEM-OF-SYSTEMS

The term System-of-Systems (SoS) is utilized to describe systems composed of multiple types of system elements where each element is an operational system in its own right. Naturally, all complex systems when decomposed result in a system composed of systems as was denoted in Figure 1-9. However, the term SoS has arisen to describe the integration of systems that have been independently designed and developed for a particular purpose, need or mission and thus can stand alone in providing those required system services. However, due to a new need (situation system), the systems are integrated in a respondent system either to respond to a real situation or to study in providing training related to a thematic situation. In some cases, an SoS can be created to provide a new sustained service, for example to merge governmental agencies or merge enterprises in order to meet new problem or opportunity situations.

To illustrate the SoS concept, consider a respondent system defined for meeting a real or thematic emergency situation that can involve a fire brigade, the police force, a military force, a medical capability, a psychological treatment capability, and so on. All of these systems have been designed and developed independently, but come together as an integrated system in meeting the needs of the emergency situation as portrayed in Figure 1-12.

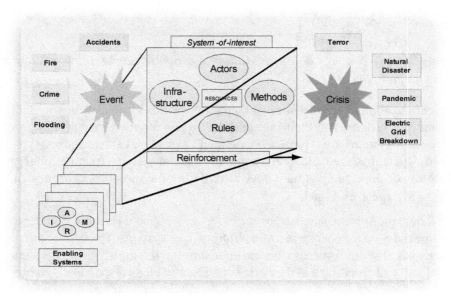

Figure 1-12: A Crises Management System of Systems

This portrayal by former course participants [Jennerholm and Stern, 2006] shows how the sustained assets of individual agencies planned to be instantiated to meet particular types of situations are brought together in a respondent system-of-systems in order to meet the needs of a crises situation. The view of the institutionalized system assets of the individual agencies as provided in this portrayal focuses upon a general view of resources. Namely, the resource assets are categorized according to Actors, Infrastructure, Methods, and Rules. This categorization is implicit in the enumeration of institutionalized assets portrayed in Table 1-1, and is used here to present an essential common denominator asset view for this purpose. As noted, the SoS needs reinforcement as a control element in order to operate as an integrated entity in responding to the crisis situation. To provide this reinforcement some form of leadership in the form of command and control is essential.

As noted earlier an SoS can also be the result of integrating the systems of several existing enterprises into a extended enterprise that is intended to provide sustained products or services over a long period of time. This can occur, for example, as the result of corporate mergers, takeovers, etc. In the public sector, it may occur when multiple governmental bodies are integrated to meet a new demand; for example, the United States Department of Homeland Security operates as an SoS in providing services based upon the integration of system assets supplied by multiple governmental agencies.

MANAGING SYSTEM CHANGE

Regardless of the enterprise, individual, group or team viewpoint concerning in-stitutionalized systems, whether they are viewed as infrastructure assets, products, or services, whether defined abstractly or concretely or at what level they exist in a recursive system hierarchy, there are three essential system related aspects associated with their life cycle management. These aspects are portrayed in the Change Model presented in Figure 1-13.

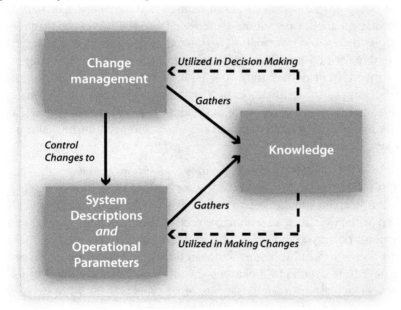

Figure 1-13: Fundamental Change Model

The adage that the only thing that is constant is change is certainly relevant to the man-made systems upon which an enterprise depends. Thus, change manage-ment is one of the most central enterprise operational functions in the life cycle management of system assets. Fundamentally, there are two types of changes that are made to system assets:

1. Structural changes – which are achieved by changes in the system description

    As noted from the point of view presented earlier, systems have descriptions; consequently deciding to make changes (i.e. transformations) on a system involves changing the description of an SOI. Structural changes can involve creating or retiring an entire system, adding or removing system elements, adding or removing services of system elements and/or redefining relationships between system elements.

2. Operational changes – which are achieved by alteration in operational parameters

    An operational change does not change the system description. It can affect the behavior of the system service(s) based upon the quantity of resources applied. For example, to operate and support multiple instances of an SOI or to provide/consume more or less resources in the form of raw material, financial support, or personnel.

    Another form of operational change is a change of mode of operation when the defined and instantiated SOI in operation provides for multiple modes. For example making a shift from normal operation mode to reduced operation mode or maintenance mode.

Key to making proper decisions about change is the gathering of pertinent data and information about operational experience and the life cycle management of systems. Such data and information is utilized in providing feedback in the form of knowledge for rational decision-making and for planning the changes as well as feed forward in providing the know-how for making prudent structural or operational changes. As significant problem and/or opportunity situations arise, the enterprise should construct respondent systems to be able to study and learn about real or thematic situations that should be taken into account in making changes in their system assets.

The ISO/IEC 15288 standard provides a variety of processes that can be utilized in fulfilling the needs of each of these essential aspects of managing system change. Further, the standard provides a basis for formulating life cycle models composed of stages that are essential to effective system change management. Each of the aspects portrayed in Figure 1-13 and the support provided by the standard provide a focus on what it means to act in terms of systems as described during this journey. In the remaining chapters of the book, this Change Model is used as a basis for explaining various aspects of systems and their life cycle management. A first set of concepts for Change Management in respect to the application of international life cycle standards for software and systems was developed by [Bendz and Lawson, 2001].

While the model portrayed in Figure 1-13 focuses upon organization/enterprise decision-making it can also serve as a conceptual model for all forms of command and control in military and/or crises management situations. The difference is related to time resolution. In acute situations, although alterations in system definitions that lead to new operative instances of the system can be created, there is most often no time available for formal descriptive changes. Thus focus is not placed upon changes of system assets, but upon changes in operational parameters of the instantiated operational system assets that are available. In any event, the assimilation of knowledge and its usage as both feedback and feed forward is vital even in these stressful situations. Experience from operations must however be used continually to evaluate the portfolio of system assets that are available. This knowledge can then be used in constructive manner in actually managing the system assets and performing change management in accordance with the model provided in Figure 1-13.

SYSTEM COMPLEXITIES

There are several ways of considering complexities in systems. As noted earlier people can have various concerns and viewpoints concerning systems and thus view systems as being simple, complex or somewhere in between. Ashby provides the following relevant exemplification of such viewpoints.

> *"... to a neurophysiologist the brain, as a feltwork of fibers and a soup of enzymes, is certainly complex: and equally the transmission of a detailed description of it would require much time. To a butcher the brain is simple, for he has to distinguish it from about thirty other "meats."*
> **W.R. Ashby** [Ashby, 1973]

Categorizing Complexity

System complexity arises in two fundamental forms as identified by Peter Senge [Senge, 1990]; namely *detail complexity* and *dynamic complexity*. Detail complexity arises from the volume of systems, system elements and defined relationships in either of the two fundamental system topologies (hierarchy or network). This complexity is related to the systems as they are; namely their static existence. Dynamic complexity, on the other hand, is related to the interrelationships that arises amongst instances of systems during their operation. That is, the expected and even unexpected behavior that actually emerges. These two forms of complexity can be directly related to system concepts presented in this chapter and can synonymously be referred to as *structural complexity* and *behavioral complexity*.

Structural and/or behavioral system complexity is related to the systems themselves as well as how the systems are perceived by people, as noted in the quotation by Ashby concerning the brain.

In relationship to systems themselves and in addition to the number elements and relationships, factors such as linearity or non-linearity in relationships, asymmetry of elements and relationships determine the degree of complexity.

In relationship to people and systems such factors as values and beliefs, interests, capabilities as well as notions and perceptions of systems are determinants of perceived complexity. As described earlier this affects how individuals and even groups view systems.

[Weaver, 1948] provided an early viewpoint by categorizing complexity into organized simplicity, organized complexity and disorganized complexity. These categories and later reflections by amongst others [Flood and Carson, 1993] and your author provide impetus for the following complexity categorization.

Organized simplicity occurs when there are a small number of essential factors and large number of less significant and/or insignificant factors. Initially a situation may seem to be complex, but on investigation the less significant and insignificant factors are taken out of the picture and the hidden simplicity is found.

Finding this simplicity is typical for scientific investigations as noted earlier in our discussion of the need to prove or disprove a scientific hypothesis. However, it is desirable to seek simplicity in all seemingly complex situations. The well-known adage KISS (Keep in Simple Stupid) reflects this viewpoint. Also, Albert Einstein also once stated: "*Make it as simple as possible, but not simpler.*" This is certainly good advice both in scientific research as well as in the engineering and life cycle management of defined physical, defined abstract and human activity systems.

Organized complexity is prevalent in defined physical and defined abstract systems where the structure of the system is organized in order to be understood and thus be amenable to scientists in describing complex behaviors as well as for structuring the engineering and life cycle management of complex systems. There is a richness that must not be over-simplified.

Disorganized complexity occurs when there are many variables that exhibit a high level of random behavior. It can also represent the product of not having adequate control over the structure of heterogeneous complex systems that have evolved due to inadequate architectural control over the system life cycle (complexity creep).

People related complexity where perception of any system fosters a feeling of complexity. In this context, humans become "observing systems". We could also relate this category to systems in which people are elements and can well contribute to organized simplicity, organized complexity or disorganized complexity. The rational or irrational behavior of individuals in particular situations is of course a vital factor in respect to complexity.

The Catastrophe Gap

Historically the rate of growth of complexity of systems has intensified over time. At the same time our ability to in some sense master (or even understand) the growing complexity has not grown at the same rate. Christer Jäderlund (a deceased Swedish system thinker) referred to the difference between the growth of system complexity and our ability to deal with complexity as the catastrophe gap as illustrated in Figure 1-14.

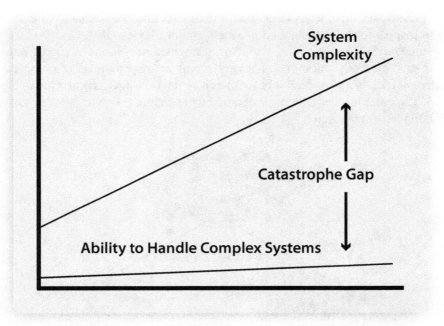

Figure 1-14: Jäderlunds Catastrophe Gap

We can take any arbitrary point in time as a starting point, for example the state of complexity in systems as of 1940 (during World War II) and then project it as some form of accelerated growth to the present day. The ability to handle the complexity has grown only slightly. So, it is not surprising that catastrophes such as the 2008 financial crises have occurred. Certainly the introduction and unprecedented growth in computer and communication technology have been driving factors in compounding system complexities. But there are many other examples of complexity contributions as well.

While many people meet system complexity with apathy or a sense of hopelessness, it is hoped that the readers of this book will arise to the task of learning how to improve their individual and collective capabilities to deal with complex systems. Quite likely, our common future will depend on progress in this area.

Complexities in Enterprises

The set of institutionalized system assets that an enterprise defines, develops or acquires and utilizes give rise to a variety of complexities. The complexities must be dealt with both by architecting the systems to minimize complexity as well as by managing the systems in a prudent manner during their life cycles.

Enterprises must learn to deal with both structural and behavioral complexity in respect to the value added products and/or services that they produce and consume, as well as complexities in their infrastructure systems including relevant processes, methods and tools. These elements become sources of complexities both individually and collectively via their interrelationships as portrayed in Figure 1-15. The complexities become magnified when dealing with the system assets of an extended enterprise.

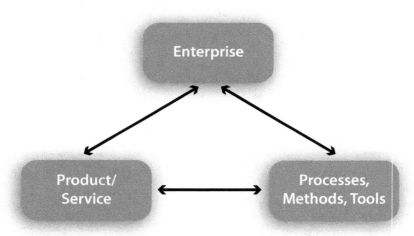

Figure 1-15: A Source of Complexity

Far too often, individuals and/or groups of people within an enterprise consider one of these elements at a time and draw conclusions as to where they believe, or for some reason would like to convince others that, their complexity problems lie. Thus they speculate that complexity problems are due to:

- Weak or inadequate product/service architecture
- Improper or ineffective enterprise structure
- Not having well defined processes or having the wrong processes
- Not having appropriate methods and/or proper tools

As a result, individuals and groups within an enterprise chase problems and opportunities from limited parochial perspectives. They may make a concerted effort to simplify their product/services in order to make it more viable to handle. Or, they concentrate on the processes to be used in their business, product/service provision and/or infrastructure of facilities, personnel, financing, and so on. They may seek out methods and tools that they believe will support their efforts. However, the easiest aspect to change is the structure of the enterprise. When in doubt – reorganize!!! Unfortunately, this is often an escape from the real problems that lie in other areas, quite often due to not having a holistic view of the complex interrelationships between all of the important elements.

The diverse interests of individuals and groups within an enterprise; namely owners and employees as well as the customers result in various forms of tension relationships as portrayed in Figure 1-16. These stakeholder tensions while absolutely essential to attaining holistic enterprises, if not handled (i.e. managed) properly lead to additional complexities [Low, 1976].

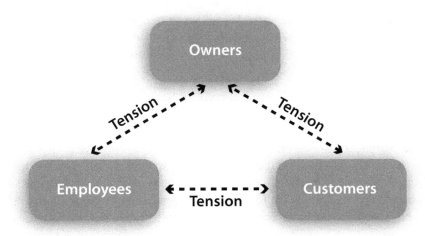

Figure 1-16: Tensions Amongst Stakeholders

The tensions become intense when large extended enterprises are involved in producing complex products and/or services. In such cases, there may not even be any clear ownership of the enterprise system thus compounding complexities.

As noted above, Einstein once stated: "*Make it as simple as possible, but not simpler.*" Certainly this is good advice in respect to the system assets of an enterprise concerning structural as well as potential behavioral complexities.

While it is possible and desirable to achieve simplicity in some types of systems, the multitude of environment related requirements placed upon enterprises and the tensions between stakeholders become constraints that most often lead to complexities in system solutions. Constraints and tensions arise due to a variety of factors including the following:

 - Laws and regulations (labor, environmental, health, etc.)
 - Patent restrictions
 - Commitments in the form of agreements and contracts
 - Product/Service historical evolution
 - Organizational related politics
 - Roles, values and norms
 - Psychological and sociological factors

Factors such as these affect the system assets of an organization and can lead to change decisions that are reflected in the structural aspects of a system. However, in the process of making structural changes in some systems, additional change requirements in the form of problems and/or opportunities arise in other systems. These changes lead to further changes thus compounding complexities, both structural and behavioral.

When problems are arising as a result of structural or behavioral complexities and/or tensions, the enterprise should establish a thematic situation and respondent system study in order to investigate the situation and use the results as a basis for Change Management. Examples of such situations to be treated by thematic systems are portrayed in Table 1-3.

A certain level of complexity is unavoidable; it is necessary. However, there are many system products and services, processes, methods and tools as well as enterprises that contain significant amounts of unnecessary complexity. One prime example of unnecessary complexity is the combined hardware and software products and services provided by the computer industry. Such complex and unstable products have lead to a multitude of compound complexities that affect individuals and their enterprises daily in the form of computer viruses, bugs, malicious attacks, and so on. To counteract these complexities, complex processes, methods and tools are developed and utilized but often leading to the compounding of complexities. A prime example of how disorganized complexity creeps in over time potentially leading to a catastrophe. In summary, MIT professor Daniel Jackson makes the following pungent observation:

> *"The question is not whether complexity can be eliminated but whether it can be tamed so that the resulting system is as simple as possible under the circumstances. The cost of simplicity may be high, but the cost of lowering the floodgates to complexity is higher."*
> **Daniel Jackson** [Jackson, 2009]

These system complexity aspects should be kept in mind on the journey through the Systems Landscape. During the journey, a variety of aids for *thinking* and *acting* in terms of systems provide useful approaches to coping with complexities of system assets, system products and system services as well as situations (real and thematic). At the end of the journey in Chapter 8, the reader will discover that organizations (as enterprises), an enterprise in an organization or an extended enterprise are themselves systems and thus have structural and behavioral properties. Viewing them is this manner will unify the knowledge attained during the journey and lead to an understanding of how to improve enterprise structures in order to attain superior enterprise behavior.

A SYSTEMS SURVIVAL KIT

In this last section of this introductory chapter in which concepts and principles have been presented informally, we formalize the concepts and principles of systems by providing concrete system semantics. The concrete semantics builds upon concretely defined concepts, concretely defined principles and the deployment of the system-coupling diagram (Figure 1-10) as a universally applicable mental model. Together these elements form a systems survival kit. That is, when understood and appreciated they will continually come to your aid individually and in groups as a means of focusing upon the essential properties of any type of system. This is the first big step in moving from mystery towards mastery as was indicated at the beginning of this chapter [Flood, 1998].

Concrete Concept Definitions

The concepts that have been introduced in this chapter are categorized and given the specific definitions provided in Table 1-4. The categories fundamental, types, topology, focus, complexity and role convey the essential properties of systems.

Table 1-4: Concrete Concepts (Categories and Definitions)

Concept Categories	Concepts	Definitions
Fundamental	Togetherness	Two or more elements are related resulting in a new whole.
	Structure	The constituent elements and their static relationship.
	Behavior	The effect produced by the elements and their dynamic element relationships in operation.
	Emergence	The predictable or unpredictable behavior occurring as the result of a system in operation.
Types	Defined Physical System	Two or more physical elements are integrated together producing a new whole.
	Defined Abstract System	Two or more abstract elements are related resulting in a new whole.
	Human Activity System	Two or more elements, at least one involving a human activity are integrated resulting in a new whole.
Topologies	Hierarchy	A level-wise structure of systems and system elements that is defined recursively.
	Network	A node and links structure of system elements and their interrelationships.
Focus	Narrow System-of-Interest (NSOI)	The system upon which focus is placed in respect to a view.
	Wider System-of-Interest (WSOI)	The systems that directly affect (including enabling) the NSOI in respect to a view.
	Environment	The context that has a direct influence upon the NSOI and WSOI.
	Wider Environment	The context that has an indirect influence upon the NSOI and WSOI.
Complexity	Organized Simplicity	There are a small number of essential factors and large number of less significant and/or insignificant factors.
	Organized	The structure is organized in order to be understood and thus be amenable for describing complex behaviors.
	Disorganized	There are many variables that exhibit a high level of random behavior. Can be due to not having adequate control over the structure of heterogeneous complex systems (complexity creep).
	People Related	Perception of the system fosters a feeling of complexity. Also, rational or irrational behavior of individuals in particular situations.

Roles	Sustained System Asset	A system that is life cycle managed and when instantiated provide system services.
	Situation System	Two or more elements become related together resulting in a problem or an opportunity. Alternatively, an objective or end state that defines a desirable situation is established.
	Respondent System	A system composed of two or more elements that are assembled in order to respond to a situation.
	Thematic System	A system that is composed for the study of possible outcomes of a postulated situation system as well as one or more respondent systems ("what if").

Universal Mental Model

The system-coupling diagram presented earlier in this chapter is repeated here as Figure 1-17. Via the system roles that are portrayed, it becomes a universal mental model for the occurrence, composition and deployment of systems. The reader should always keep this model in mind, as we shall continually return to it during the journey.

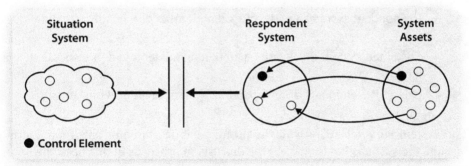

Figure 1-17: Universal Mental Model of Systems

System Principles

Building upon the concrete definitions of concepts and the utilization of the system-coupling diagram as a universal mental model, we can now express concrete principles that establish the following system rules (truths to abide by):

- All systems are composed of two or more elements that constitute togetherness
- Systems are composed of structural elements or behavior elements
- Defined elements and relationships can be abstract, physical or human activities
- Systems are organized as a hierarchy or a network
- Bounding of systems in respect to views are defined by a Narrow System Of Interest, its Wider System Of Interest, their Environment and Wider Environment
- Complexity can be reduced by the identification of essential factors (concepts and principles)
- Complexity is addressed by proper organization in describing complex behaviors
- Complexity rises when systems are disorganized resulting random behavior
- People have various perceptions of complexity as well as potentially participating in a system resulting in the addition of complexity
- Situation systems result from (problems or opportunities) or from defined objectives in the form of end states
- Respondent systems are developed and utilized to handle situation systems
- Sustained system assets are instantiated and deployed in respondent systems
- One of the elements of a respondent system must provide control

The system survival kit provides a strong basis for moving on in our journey through the systems landscape. As the discussion proceeds in this book the reader will discover that one of the prime advantages of learning to think and act in terms of systems is the potential for new inventiveness (fantasy) that arises due to understanding the interrelationships of multiple systems and the possibilities that arise by combining systems into a new system idea.

It is soon time to get under way. But first verify your knowledge of the system concepts presented in this introductory chapter.

KNOWLEDGE VERIFICATION

Each chapter in this book contains a knowledge verification section that provides questions and exercises aimed at verifying the knowledge of the reader. In answering the questions and doing the exercises, it can be useful for small groups of colleagues to delve into their own experiences and points of view in order to facilitate knowledge acquisition via discussion and dialogue.

1. Identify structural and behavioral properties in disciplines with which you are familiar.

2. Identify some natural systems and several defined physical, defined abstract and human activity man-made systems.

3. Identify several examples of systems organized in hierarchical and network topologies.

4. Using the examples produced in (2 and/or 3), identify various actors (individuals or groups) that may view the systems according to different viewpoints.

5. Discuss your perspective with others as to the existence of systems. That is, do they actually exist or are they simply abstract descriptions? Support your reasoning as to why they are real or why they do not exist.

6. Identify systems that are important for an enterprise with which you are familiar.

7. Describe the need for, services provided and delivered effect of some systems upon which you rely.

8. Identify the system elements of a familiar system-of-interest.

9. What is meant by the emergent behavior of a system?

10. Identify the relationships between the system elements identified in (8).

11. What is meant by the recursive decomposition of a hierarchical system?

12. Decompose one of more of the system elements identified in (8) into a separate system-of-interest composed of their own system elements

13. When is recursive decomposition of a hierarchical system-of-interest terminated (stopped)

14. Compose several examples of systems that incorporate multiple types of system elements

15. Identify examples of sustained, situation, respondent and thematic systems

16. What is meant by the terms extended enterprise and system-of-systems

17. How are changes, systems descriptions and knowledge treated in enterprises with which you are familiar.

18. Describe and exemplify the difference between a change in operational parameters and a change in system description.

19. Identify cases where organized simplicity, organized complexity, disorganized complexity and people related complexity are apparent.

20. Describe some of your own experiences with structural and behavioral system complexities as a provider and/or user of system products/services. Consider how the identified complexities compound by affecting the enterprises, and/or its processes, methods and tools

21. Identify the impact of various forms of tensions upon the systems in an enterprise with which you are familiar

22. Verify that the systems survival kit is indeed universal by trying to find examples of open man-made systems that do not fit into the framework of the kit.

Note 1-1: There are a variety of definitions of the terms organization and enterprise. In the ISO/IEC 15288:2002 standard the following definitions are provided:

Organization – a group of people and facilities with an arrangement of responsibilities, authorities, and relationships. Definition taken from [ISO 9000:2000]

Enterprise – that part of an organization with responsibility to acquire and to supply products and services according to agreements.

Note: An organization may be involved in several enterprises and an enterprise may involve one or more organizations.

The note provided in the standard indicates that enterprises can be singular, but can span a variety of organizations, thus becoming extended enterprises. However, the definition of an enterprise provided by the standard while correct is somewhat restrictive in that it relates mostly to trading of system products and/or services based upon agreements. As the journey unfolds in the book, the reader will observe a more general perspective concerning enterprises. [Rouse, 2005] captures a more general view of enterprises by stating:

"An enterprise is a goal-directed organization of resources – human, information, financial, and physical – and activities, usually of significant operational scope, complication, risk, and duration. Enterprises can range from corporations, to supply chains, to markets, to governments, to economies."

The view taken in this book is that an enterprise is any type of endeavor that leads to the achievement of purpose, goals, and missions, including the acquisition and/or supply of products and services. Obviously an enterprise also has an organization. Since an organization also exists for a purpose, has goals and works in fulfilling missions, it is also an enterprise. Thus, we can for all practical purposes view the terms organization and enterprise as being equivalent.

Chapter 2
Thinking in Terms of Systems

Think before you act; then rethink before acting again

An enterprise that operates effectively learns from deploying its institutionalized system assets (the operation of its systems), from the study of real or thematic situation and respondent systems relating to problem and opportunity situations as well as learning from managing the life cycles of its systems. The knowledge gained, as portrayed in Change Model of Figure 1-13, is feedback and feed forward for the enterprise and represents vital human intellectual capital for future change management. In order to achieve the culture of such a "learning organization" that can address system complexities it is essential that individuals and groups (projects and teams) develop the capability to *think* as well as to *act* in terms of systems. So, in this chapter and the following chapter we look closer at what this actually means.

SYSTEMS THINKING

"Systems thinking is a process of discovery and diagnosis – an inquiry into the governing processes underlying the problems we face and the opportunities we have."
[based upon **Senge et.al.**, 1994]

As noted in the introductory chapter, modern systems thinking evolved during the 20th century, via multiple contributions into a somewhat more understood discipline. Building upon the pioneering contributions of Ludwig von Bertalanffy in the 1920s,

Jay Forrester, Russel Ackoff, Ross Ashby, Staffard Beer, Wes Churchman, Peter Checkland, Peter Senge and others have made important contributions to systems thinking in the latter half of the 20th century.

Systems Thinking can be considered to be an essential part of the discipline of Systems Science as was inferred in Figure 1-1 when the unification of disciplines was presented. Thus, like other scientific endeavors, it is related to observing, finding important properties and describing. In contrast to more specific scientific disciplines striving to establish principles and theory, Systems Science has evolved from contributions in multiple disciplines. The disciplines of Operations Research and Decision Analysis, also being generally applicable in and amongst multiple disciplines can be considered to be sub-disciplines within System Science.

In 1955, the biologist von Bertanlaffy together with an economist (K.E. Bolton), a physiologist (R.W. Gerard) and a mathematician (A. Rapoport) founded an organization with the goal to advance Systems Science by the establishment of a forum for General Systems Theory that was based upon the following aims:

1. To investigate the isomorphy of concepts, laws, and models in various fields, and to help in useful transfers from one field to another;
2. To encourage the development of adequate theoretical models in areas which lack them;
3. To eliminate the duplication of theoretical efforts in different fields;
4. To promote the unity of science through improving the communication between specialists.

This multidisciplinary team succeeded in the building of a society that has provided an important public forum for contributors; however, actual development in the field has not met these specific aims. There is still diversity in concepts, models, etc. as there has not evolved a single set of concepts, principles and theory that is universally accepted. Thus, duplication of theoretical and practical efforts is still ongoing and is required in the quest to learn more about the nature of systems thinking and its application. The reference to a unity of science is certainly too restrictive in that both scientific and non-scientific disciplines have contributed to and benefited from the system thinking concepts that have evolved from multiple sources. Thus, it is the unification of disciplines in respect to structures and behaviors, as pointed to in the previous chapter, that is characteristic for establishing both theory and practice in the field of systems thinking.

Systems thinking, also referred to as *systemic thinking*, was popularized during the 1990s due to the contributions of Peter Checkland [Checkland, 1993] and Peter Senge [Senge, 1990], [Senge et.al., 1994]. These two seminal contributions and others have been described, criticized, and contrasted by Robert Flood [Flood, 1999].

Essential aspects of systems thinking are presented in this book that reflect various approaches to the topic; however, a thorough discussion of all contributions to systems thinking is not within the scope of this book. The reader is referred to the referenced books as well as a Web search on "systems and systemic thinking" as starting points for their own deeper investigation of the discipline. System thinking is, after all, a process of inquiry.

A major thrust of systems thinking is in identifying, observing and understanding the complex emergent behaviors arising from the dynamic interactions of multiple systems in operation. As a result, decisions may be made to alter, eliminate or introduce one or more systems. The ability to act in terms of systems (that is effectively make structural system changes) is not covered in the literature on systems thinking, but it is exactly in this area that systems engineering naturally complements systems thinking as described in the introductory chapter. In particular, via utilizing system engineering related processes for system life cycle management, managed changes in respect to system descriptions and operational parameters (as portrayed in the Change Model of Figure 1-13) can be effectively executed, verified and validated during the life cycle of systems-of-interest. These action related aspects are described in Chapter 3.

Peter Senge describes systems thinking as:

- a discipline for seeing wholes

- a framework for seeing interrelationships, for seeing patterns of change rather than static "snapshots

- a set of general principles-distilled over the course of the twentieth century, spanning fields as diverse as physical and social sciences, engineering and management

- also a specific set of tools and techniques

According to Senge and his colleagues, a good systems thinker, particularly in an organizational setting, is someone who can see four levels operating simultaneously: events, patterns of behavior, systems, and mental models.

In seeing wholes, interrelationships and patterns of change in systems, it is important to be aware of the dangers of utilizing simplifying counterproductive approaches. The problems associated with scientific *reductionism* have been indicated earlier. Another danger that stands in the way of seeing wholes, interrelationships and patterns of change is the restriction to making generalizations (inferences) based upon singular *relationships*; for example A *causes* B. It is vital to expand the view of causal relationships to the multiplicity of relationships that exist in the dynamic interaction of non-trivial systems; that is A *causes* B, which in turn *causes* C and D which in turn *causes* … and so on. Further, the relationships may not be so directly coupled in a causal sense, in which case the use of the term *influence* is more appropriate to describe the relationship.

According to another perspective, Checkland argues that system thinking is founded upon two pairs of ideas; namely, *emergence and hierarchy* and *communication and control*. In Chapter 1, emergent behavior and the hierarchical decomposition of systems were described as fundamental system concepts. Communication is generalized in describing relationships between elements in a system as well as between the system and its environment. Thus, it can involve the exchange of materials, energy, or information. Control within a system is based upon the availability of information that measures relevant parameters in on-going processes. Control in both natural as well as man made systems has been described via the system theory of *cybernetics* in which feedback mechanisms provide for regulation. In this respect, the Change Model introduced in Figure 1-13 is an example of control exercised by humans in decision-making. These control aspects will be explored in later chapters.

In summary, the principles of system thinking have evolved as a result of observing common holistic aspects of systems in diverse fields of endeavor. It is founded upon an understanding that there are common relationships between systems in nature and in and amongst man-made systems that are useful to understand and exploit. In fact, systems thinking, as an essential part of systems science, and systems engineering both are major contributors in achieving a unification of disciplines in respect to systems.

HARD AND SOFT SYSTEMS

The word system for many people carries with it a connotation that it is something that exists in the world and is composed of set of interacting systems elements. This view, and the counter-view that man-made systems exist as descriptions, was introduced in Chapter 1. In any event, a dominant view for man-made systems is that they are engineered to achieve a purpose, goal or mission that is of interest. The field of systems engineering which parallels to a large extent the developments of systems thinking in the latter half of the 20th century has evolved to deal with such engineering challenges. While early focus in systems engineering was placed upon large-scale physical systems, the field has evolved to include organizational, management and human factor issues from a total life cycle perspective [Arnold and Lawson, 2004]. Historically, there have been attempts to apply the methodology of systems engineering to less structured social and political systems that often fall short of gaining the necessary insight needed for system improvement.

Checkland [Checkland, 1993], starting himself from a systems engineering perspective, successively observed the problems in applying systems engineering to the more fuzzy ill-defined problems found in the social and political arenas. Thus he introduced a distinction between hard systems and soft systems.

Hard systems of the world are characterized by the ability to define purpose, goals and missions that can be addressed via engineering methodologies in attempting to in some sense "optimize" a solution.

Soft systems of the world are characterized by extremely complex, problematical and often mysterious phenomenon for which concrete goals cannot be established and which require learning in order to make improvement. Such systems are not limited to the social and political arenas and also exist within and amongst enterprises where complex, often ill-defined, patterns of behavior are observed that are limiting the enterprise ability to improve.

Recognizing this important difference, Checkland points to the fact that a process of inquiry that itself can be organized into a learning system is the most appropriate approach for analyzing and learning about soft systems in which human activities exist as elements.

Systems thinking is useful in both analyzing and gaining insight into the problems and opportunities associated with hard systems and soft systems as well as systems containing both types of elements.

MODELS AND MODELING

A number of methodologies, tools, models, languages, and techniques have been developed in relationship to systems thinking that assist in the fundamental aspects of seeing wholes, interrelationships, and patterns of change. The work of Professor Jay Forrester at MIT during the 1950 and 60s led to the development of the DYNAMO Simulation Language that was one of the earliest computer-based simulation languages and provided a methodology and tool for studying the complexities of dynamic system relationships [Forrester, 1975]. Forrester was also early in pointing to the universality of systems thinking in multiple disciplines and worked on both hard and soft systems. Other programming languages for simulating complex systems arose during the 1960s such as SIMSCRIPT [Markowitz, 1979] and SIMULA [Dahl, et. al., 1970].

Gaining insight into complex systems typically involves the development of one or more models that attempt to capture some aspects of the structure and/or potential behavior of a system. For example, in natural systems, models of weather systems are constructed based upon measurements and known patterns of hydrological behavior. These models are constantly analyzed dynamically in order to provide weather prognosis. Mathematical models of structures and behaviors of natural phenomenon based upon the laws of physics, biology or chemistry are constructed to capture some relationships between physical elements. The models may be expressed manually at the paper and pencil level or expressed in some form of language that provides a basis for computer-based simulation.

Models for man-made systems are useful for systems thinking related to hard or soft systems. Models of systems of the defined physical variety are based upon mathematical formulae defining elements and relationships. The Mathematica and MATLAB products have become important tools in building and analyzing models for natural systems as well as designed physical systems. [Fritzson, 2004] has provided a comprehensive book concerning modeling and simulation of physical systems based upon an object-oriented language called Modelica-2.

As an example of a model of a defined physical system consider an automatic production system for producing some type of physical products. The model can, for example capture the physical production equipment in the form of work cells and buffers. The processing of raw and semi-processed materials during the production process can be described by the use of formulae of probabilistic distributions, for example, Gaussian or Poisson. These models are used both for gaining insight and for verifying that the model portrays the reality of the production process.

Models for defined abstract systems can involve similar analysis where hypothesis concerning processing capacities can be used in gaining insight into a set of functions and/or capabilities and their interrelationships. Mathematical formulae may also be used in describing the relationships as to rate of processing when the abstract system will eventually be developed into a physical system.

Models of human activity systems can be constructed in order to gain insight into real or thematic situation and/or respondent systems related to some problem or opportunity. Human activity models composed of functions and/or capabilities and interrelationships can also be utilized to capture sets of processes and/or procedures that are to be carried out by humans.

Since models are always an abstraction of reality and not reality, it can be concluded that ***all models are wrong; however some of them are useful*** [Box & Draper, 1987]. Boardman and Sauser [Boardman and Sauser, 2008] point to the following useful conclusions about models. No model should be built unless we know:

– What we are looking at.
– Why we are looking at it.
– From where (which standpoint) we are looking at, and
– What it is we believe we can see better because we have the model.

  The last thing of significance in building a model, in our consideration is the *how*.

Independent of what form of models are built, be they quantitative or qualitative, that which is most important from the systems thinking point of view is a well founded modeling process. Thus, as noted above, it involves as Peter Senge points out, establishing a holistic view and the ability to determine patterns of action rather than static snapshots of a situation. Also, as Peter Checkland points out that models provide a basis for learning about system situations.

PARADOX

System situations are often paradoxical and contain contradictions (phenomena that can be viewed both as being true and as being false). A classic example is: This sentence is a *lie*. If the sentence is true, then it is a lie; however, it if it is false, how can it be true. When uncovering system situations, the system thinker must be aware of paradoxical situations. Boardman and Sauser relate the following concerning systems thinking and paradoxes.

> *"A paradox is an apparent contradiction; however, things are not always as they seem. A paradox can be explained, but only by seeking wisdom from above; for the systems person this means looking upwards and outward, not just down and in. Paradoxical thinking is systems thinking at its best."*

Paradoxes where statements claiming to be true and at the same time contradicting each other cause tension that is sometimes hard to accept for the parties involved. At the same time these conflicts when examined by models that lead to new thinking can provide a change of mindsets that lead to deeper understanding. Let us consider some examples.

Control Paradox – Command and Control are utilized in order to assure orderliness and conformance to some form of strategic direction (purpose, goal or mission). However, in order to foster innovation, creativity and a sense of self-awareness you must not have Command and Control.

Customer Paradox – Listening to your customers is vital for sustaining productivity and profit. However, to take advantage of disruptive technology that can provide new sources of products and profits, you must not listen to your customers.

Diversity Paradox – In order for teams (including projects, missions, task forces, etc.) to succeed there is a need for sameness as well as differentiation. Sameness must exist in order to provide esprit de core and a feeling of common goal togetherness. However, differentiation in respect to member capabilities and skills is essential for providing team success.

Software Paradox – Computer software due to its fundamental nature leads to the paradoxical situation that software systems should be planned while at the same time omnipresent programmer creativity demands that they not be planned. This is the central area of debate concerning planned versus agile programming.

The system thinker, when considering multiple system perspectives must look upwards and outwards as well as down and in and must consider the NSOI (narrow system-of-interest), the WSOI (wider system-of-interest) as well the Environment and Wider Environment as portrayed in Figure 1-5. Thus, understanding paradoxical situations and the tensions they create as well as the drive to look at holistic aspects of a problem or opportunity should be kept in mind as we look at some of approaches to modeling (describing) system situations. Some of the modeling approaches can be utilized for hard and soft systems; whereas, others are more dedicated to modeling hard, respective soft systems. However, let us first examine the systems and relationships to be described.

DESCRIBING SYSTEM SITUATIONS

There are a variety of approaches to quantitative and qualitative modeling of systemic situations. The approaches provide some type of perspective that helps build insight into the broader aspects of systems. The qualitative models can consist of natural language prose text, structured text, pictorial portrayals and/or graphical representations of essential properties of a system related problem or opportunity. All forms of qualitative models are aimed at relating a "system story" that provides useful insight. Various forms of story telling via models will be considered in the following discussions but first let us reconsider the importance of the system coupling diagram paradigm.

System Focus

The identification of problems or opportunities is most often related to relationships between elements of multiple systems. Thus, let us once again consider the system coupling diagram mental model in this context as portrayed in Figure 2-1.

The problem or opportunity may arise due to elements and relationships that have evolved in a situation and are bounded in the narrow NSOI Situation System. However, in looking upward and outward, the boundary of the problem or opportunity changes due to its existence in a Wider System-of-Interest (WSOI), the Narrow Environment and even Wider Environment. Naturally, problem or opportunity situations arise in Respondent Systems as well as in relationship to System Assets that need to be described. Also, in these cases, the WSOI, the Narrow Environment and Wider Environment must be taken into account.

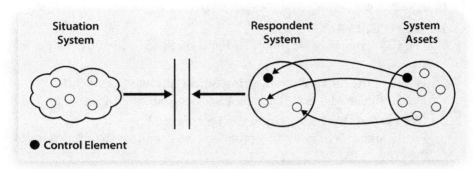

Figure 2-1: Where is the Problem or Opportunity?

Since a Respondent System is created to handle a Situation System, the coupling of these two is a ripe area for description. The elements of these two systems interact leading to the resolution of problems or opportunities, but also possibly leading to new situations. It is most important to consider a wider context when focusing on the problem including the properties of sustained system assets from which a Respondent System is constructed. So, it is important to identify elemental relationships in the NSOI, the WSOI, the Environment and the Wider Environment.

Finding the Root Causes

Finding sources of problems is, of course, a prerequisite for modeling. Many times the problems are not obvious and can be the result of some form of paradoxical situation.

Senge et.al., [1994] describe a method based upon textual representation that has proven to be useful for both individuals as well as for groups working together to identify roots causes of problems. The method, referred to as the Five Whys, provides a useful starting point to establish guiding thoughts about root causes (what is affecting what and why). In essence, the method looks further and further outward from a problem that is considered to be troublesome.

The method is best carried out using a flipchart, paper, markers, self-sticking notes where someone is identified to write everything down. Selecting the first Why (usually resulting in three or four starting points) asks the question: "Why is such-and-such taking place?"

To illustrate the use of the Five Why method, we will go through some conclusions drawn by Margaretha Ericsson [2006]. In a project by this former course participant, she examined the root causes of why Customer Product Information (CPI) for software related product is Inadequate, Late and Expensive to Develop. The following two examples are the result studying documentation related problems of a well-known telecommunication company.

1. **Why is CPI late?**
   The CPI writers are waiting for a Function Specification from a Designer.

2. **Why are the CPI writers waiting for a Function Specification?**
   The Designer who is to write the FS is delayed.

3. **Why is the Designer writing the FS delayed?**
   The Designer is busy writing program code for the system functions of the FS.

4. **Why is the Designer writing code for the system function?**
   The Designer is responsible for the program code.

5. **Why then is the Designer not writing the FS?**
   The Designer describes the system functions in the FS after coding and testing.

Certainly, the software paradox where planning must exist and must not exist in order to promote creativity is observable in this example. Now consider a related problem that is to a large extent a consequence of the previous situation.

1. **Why is the CPI work time-consuming (and getting expensive)?**
   The CPI holds many documents and pages.

2. **Why is it many documents and pages?**
   The CPI has grown organically with the system, without deliberate structure.

3. **Why has the CPI grown organically?**
   Nobody in the project has had the time or mandate to create an information structure.

4. **Why is there no information structure?**
   No single person has the full picture of the user/customer needs and the System-of-Interest.

5. **Why is there no understanding for user/customer needs?**
   There is no feedback from the users/customers.

This method based upon structured text provides a useful starting point for analyzing a variety of complex situations. We now consider various forms of graphical and pictorial representation of system situations.

Influence Diagrams

Describing what influences what is fundamental to all of the descriptive approaches. One of the must straightforward methods of modeling in this regard is the Influence Diagram. While several people contributed, influence diagrams were popularized by Stanford University Management Science Professor Ronald Howard as a part of his work in the field of Decision Analysis [Howard, 1960 and Howard and Matheson, 1984].

An influence diagram is a simple visual representation of a decision problem. Influence diagrams offer an intuitive way to identify and display the essential elements, including decisions, uncertainties, and objectives, and how they influence each other. They provide a complementary approach to decision trees (described in Chapter 5). The diagrams are composed of nodes of various shapes and arrows with the meaning indicated in Table 2-1.

Table 2-1: The Symbols of Influence Diagrams

▭	A decision is a variable that you, as the decision maker, have the power to control.
◯	A chance variable is uncertain and you cannot control it directly.
⬡	An objective variable is a quantitative criterion that you are trying to maximize (or minimize).
▢	A general variable is a deterministic function of the quantities it depends on.
↗	An arrow denotes an influence. A influences B means that knowing A would directly affect our belief or expectation about the value of B. An influence expresses knowledge about relevance. It does not necessarily imply a causal relation, or a flow of material, data, or money.

To illustrate the usage of influence diagrams, consider the example provided in Figure 2-2. This simple influence diagram shows how decisions about the marketing budget and product price influence expectations about its uncertain market size and market share as well as costs and revenues. The market size and market share, in turn, influence unit sales that influences costs and revenues, which influences the overall profit.

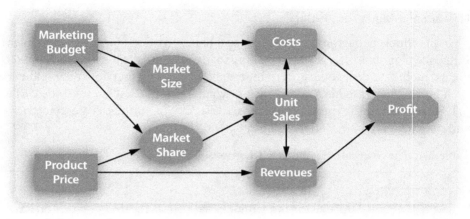

Figure 2-2: An Influence Diagram

There are tools available to support the development and utilization of influence diagrams. For example, see www.lumina.com.

Links, Loops and Delays

Peter Senge, while a student of Jay Forrester at MIT, in the 1980s developed a "systems thinking language" based upon the primitive notions of links, loops and delays. The language permits the construction of viewpoint models as representations of multiple causal relationships in interacting systems that result in behavioral patterns. In this regard, it is similar to influence diagrams, but as we shall observe, the models are based upon the two fundamental notions of *growth* and *limits*. As an example, consider the links, loops and delay representation of a problem encountered by a service provider in Figure 2-3.

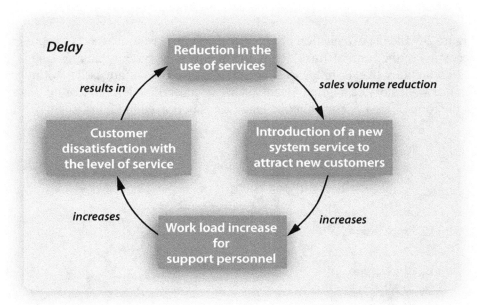

Figure 2-3: Links, Loops and Delay Example of a Service Provider Decision

Here four situations or actions related to various systems that are in operation in an enterprise can be observed. These actions and situations are collected into a pattern of behavior that can be used to explain phenomenon that the enterprise is experiencing.

In this system situation, in an attempt to attract new customers the service provider decided to introduce a new system service. As a result of this introduction, there was a workload increase for support personnel. As a result of the workload increase, customer dissatisfaction with the general level of service increased. Finally, after a delay of some period of time, customers gave up and left the service provider resulting in a reduction in the use of services. Not at all what the service provider had expected from the introduction of the new service. At this point with a declining customer base, the unwise decision may be to introduce even further new system services in attempting to increase the customer base that most likely will lead to further degradation. However, it is important to note that due to some customers leaving the service provider, the workload of support personnel decreases and assuming that the personnel resources are still in place, customer satisfaction may increase again after a delay.

Telling a Systems Story

Figure 2-3 and the explanation just provided point to a central aspect of capturing scenarios in the form of links, loops and delays. The system thinking language representations, like the influence diagrams discussed above, are used to formulate an important story of multiple relationships. A general model for creating a story in the form of the systems thinking language is portrayed in Figure 2-4.

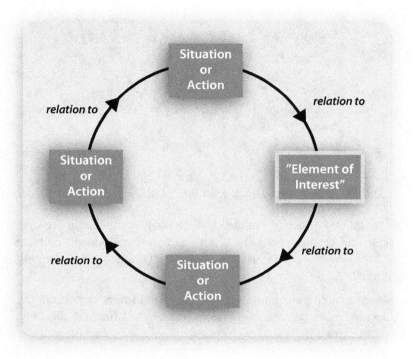

Figure 2-4: General Structure of a Systems Thinking Story

There may be more or less elements in the story, but at least two elements must be described. The following steps to telling system stories are recommended by Senge and his colleagues:

1. Start anywhere. Pick the element, for instance, of most immediate concern. Do not explain why this is happening – yet.

2. Any element may go up or down at various points in time. What has the element been doing at this moment? Try out language which describes the movement – *goes down … improves … deteriorates … increases … decreases … rises … falls … soars .. drops.*

3. Describe the impact that this movement produced on the next element. For example *as one element is going down efforts to improve another element goes up.*

4. Continue the story back to the starting place. Use phrases that show causal interrelationship: "This in turn, causes ... "or"... which influences ... "or" then adversely affects ...

5. Try not to tell the story in cut-and-dried mechanistic fashion. Make it come alive by adding illustrations and short anecdotes so others know exactly what you mean.

A natural language such as English, being linear allows us to talk about one step at a time. In complex systems multiple events happen simultaneously and thus are difficult to capture. Telling a story of multiple elements and relationships in links, loops and delay language form helps in recognizing system behavior and in developing a sense of timing.

System Archetypes

Building upon these simple notions, patterns of links, loops and delays structures have been developed to explain frequently occurring patterns of behavior in system interactions. These patterns are called *system archetypes* and include growth (reinforcing) loops (rolling snowball effect of compounding positive or negative growth), and limiting (balancing) loops (i.e. limiting factors that can break trends of reinforcing behaviors) as illustrated in Figure 2-5.

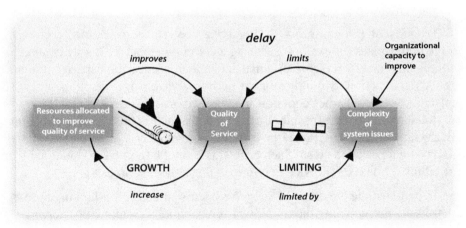

Figure 2-5: Reinforcing and Balancing Loops Example (Limits on Quality Improvement)

The story told in this figure is that in striving to provide improved quality of service, additional resources are allocated which as noted in the *reinforcing (growth) loop* on the left results in improvements in the quality of service. This improvement may continue by allocating even more resources. Eventually, there is a limiting factor denoted in the *balancing (limiting) loop* on the right that is related to the organizational capacity to improve due to the inherent complexity in system issues. Thus, after some delay in time, despite efforts to improve quality of service, a limit is reached.

The examples in Figure 2-3 and 2-5 illustrate Senge's system thinking language as well as the importance of observing complex behavioral patterns that result from changes in structures (a new service) respectively in operational parameters (increasing resources). A variety of additional system archetypes are built upon connecting reinforcing and balancing loops in various combinations to explain complex patterns of behavior as described in [Senge et..al., 1994]. The following are some examples of useful archetypes:

Fixes that backfire – A problem symptom exists for which a fix is made in order to limit the effect of the problem. However, after a time delay, the fix actually results in growing negative consequences thus compounding the problem. This is a truly paradoxical situation.

Shifting the burden - A problem symptom is connected by two balancing loops, one addressing quick fixes to the symptoms of a problem and another addressing the correction of the fundamental problem. As with fixes that backfire, there are growing negative consequences of the problem. Reinforcing loops can evolve related for example, to providing rewards for heroic behavior in quick fixes leading to addiction, that is, a dependence on heroics thus reducing the ability to treat the fundamental problem.

Tragedy of the commons - This archetype relates to people or groups of people (enterprises for example) sharing a common resource. It may be a natural resource, human effort (service), financial capital, production capacity or market size. When in action, two indicators of performance change simultaneously. The total activity, using up the "common" resource rises robustly. However the gain an individual or group feels from their effort-gain per action hits a peak and begins to fall. Eventually (after a delay) if the dynamic utilization continues the total activity will also hit a peak and crash. Such problems cannot be solved in isolation by one party it must involve fellow competitors that use the resource.

To demonstrate the usage of Senge's systems language for thinking in respect to hard systems, consider the tragedy of the commons system story portrayed in Figure 2-6.

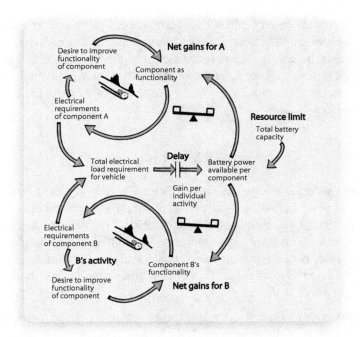

Figure 2-6: Tragedy of Commons Example: Vehicle Battery Capacity

In this example multiple components utilize a single battery source where there is a load capacity that is a shared single resource (Resource limit). Two groups of design teams A and B working independently wish to continually improve the functionality of the component they provide. They both work in their own growth loops and when their components are to be integrated using the common power supply, the electrical requirements are considered. If both A and B continue to increase the functionality and the needed electrical load requirements, they will reach limits due to the vehicle's battery capacity. However, as indicated this comes after a delay. That is the limit is experienced when the total load capacity of the battery is exceeded. The only way out of this situation is through negotiation between the groups.

These and other archetypes are associated with many common system related real or thematic situations in the form of opportunities and problems facing organizations and their enterprises. A very useful Archetype Family Tree showing relationships between how various archetypes can be used in a variety of situations is provided in [Senge et..al., 1994]. Many system and systemic thinking related websites contain further examples of the use of archetypes based upon Senge's systems thinking language. One of the most comprehensive and useful is www. systems-thinking.org which is managed by [Bellinger, 2004].

Senge refers to Systems Thinking as "The Fifth Discipline". It supports other disciplines of Personal Mastery, Mental Models, Shared Vision, and Team Learning as the essential prerequisites for achieving a learning organization. In fact, all of these disciplines contribute to the Knowledge aspect of the Change Model portrayed in Figure 1-13. Thus, further consideration of these important contributions by Senge is presented in the chapter addressing the properties of Knowledge (Chapter 7).

Rich Pictures

Rich pictures were developed as part of Peter Checkland's Soft Systems Methodology (described below) for gathering information about a complex situation [Checkland, 1981, Checkland and Scholes, 1990]. The idea of using drawings or pictures to think about issues is common to several problem solving or creative thinking methods because our intuitive consciousness communicates more easily in impressions and symbols than in words. Thus, drawings can both evoke and record insight into a situation.

Rich picture models are drawn at the pre-analysis stage, before you know clearly which parts of the situation should best be regarded as behavioral (due to an on-going process) and which is best regarded as structural. Rich picture models (i.e. situation summaries) are used to depict complicated situations. They are an attempt to encapsulate the real situation through a no- holds-barred, cartoon representation of layout, connections, relationships, influences, cause-and-effect, and so on. As well as these objective notions, rich pictures should depict subjective elements such as character and characteristics, points of view and prejudices, spirit and human nature. There are no specific conventions or rules related to rich pictures. Typical elements of a rich picture model can include pictorial symbols, keywords, cartoons, sketches, symbols and titles.

To illustrate Rich pictures, Figure 2-7 identifies the "situation" that a Toy Manufacturer faces in respect to toy safety. The aspects portrayed reflect aspects of the Wider System of Interest (WSOI) as well as the Environment and Wider Environment. Such a rich picture could well have been developed in order to explore the complexities that a toy manufacturer must deal with. In the picture you will find factors (elements) that must be taken into account.

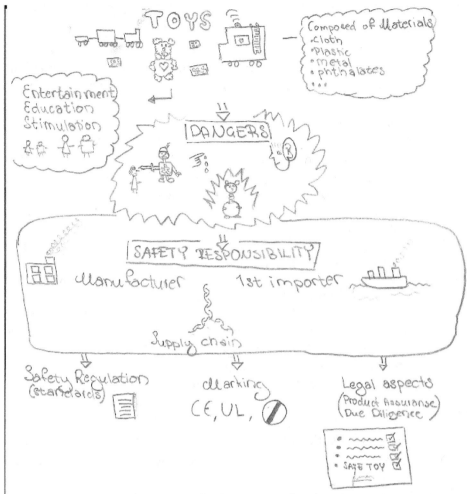

Figure 2-7: Toy Safety Situation for a Toy Manufacturer

The following guidelines are useful to consider in creating rich picture models.

1. A rich picture is an attempt to assemble everything that might be relevant to a complex situation. You should somehow represent every observation that occurs to you or that you gleaned from your initial survey.

2. Fall back on words only where ideas fail you for a sketch that encapsulates your meaning.

3. You should not seek to impose any style or structure on your picture. Place the elements on your sheet wherever your instinct prompts. At a later stage you may find that the placement itself has a message for you.

4. If you 'don't know where to begin', then the following sequence may help to get you started:

a. first look for the elements of structure in the situation (these are the parts of the situation that change relatively slowly over time and are relatively stable, the people, the set-ups, the command hierarchy, perhaps);

b. next look for elements of process within the situation (these are the things that are in a state of change: the activities that are going on (i.e. behavior));

c. then look for the ways in which the structure and the processes interact. Doing this will give you an idea of the climate of the situation. That is, the ways in which the structure and the processes relate to each other.

d. Avoid thinking in systems terms. That is, using ideas like: 'Well, the situation is made up of a marketing system and a production system and a quality control system'. There are two reasons for this. The first is that the word 'system' implies organized interconnections and it may be precisely the absence of such organized interconnectedness that lies at the heart of the matter: therefore, by assuming its existence (by the use of the word system) you may be missing the point. Note, however, that this does not mean that there won't be some sort of link or connection between your graphics, as mentioned above. The second reason is that doing so will channel you down a particular line of thought, namely the search for ways of making these systems more efficient.

e. Make sure that your picture includes not only the factual data about the situation, but also the subjective information.

f. Look at the social roles that are regarded within the situation as meaningful by those involved, and look at the kinds of behavior expected from people in those roles. If you see any conflicts, indicate them.

g. Finally you may include yourself in the picture. Make sure that your roles and relationships in the situation are clear. Remember that you are not an objective observer, but someone with a set of values, beliefs and norms that color your perceptions.

Systemigrams

Inspired by Checkland's work, John Boardman developed a version of the Soft System Methodology (described below) in which he created a form of modeling called Systemigrams [Boardman and Sauser, 2008]. According to Boardman, Systemigrams are based on a complete respect of the totality of prose, believing that its richness deserves to find graphical expression, and in that graphical expression inspires further detailed grammatical exposition leading to more detailed graphical description.

The development of Systemigrams has taken place over a period of 20 years in academic, industrial and governmental settings. Emphasis is based upon the ability to translate prose (text) into graphical form according to the following construction process:

1. To faithfully interpret the original structured text as a diagram in such a way that with little or no tuition the original author, at the very least, would be able to perceive his or her writings, and additionally, meanings.

2. To create a diagram that is a system, or could at least be considered a system in its own right. Thus, if the original structured text could be considered a system, then its faithful interpretation as a new object should be systemic, but with features not possible with prose alone, but quite amenable as a graphic (or picture).

3. To ensure not only compatibility between the graphic object and structured text, but also synergy so that both objects could evolve into more potent instances, capable of improved dissemination and community building, development and mobilization. This is not a case of either-or but both.

Boardman and Sauser provide many examples of how Systemigrams are constructed in order to reflect situations that are or could be expressed in text. To illustrate this easy to use but powerful means of representation, let us consider a Systemigram provided by a former course participant Ivonne Donate [2009]. In her project, she used Systemigrams to represent the relationship between technology levels from the TRL (Technology Readiness Levels) concept.

The TRL concept was developed in early 1970's by NASA when trying to determine a way to minimize the risk of new technologies to be used in its space program. In the mid-90s, the concept of measuring technology maturity caught the interest of other developers among which was the Department of Defense (DoD) who adopted the TRLs as defined by NASA. Table 2-1 shows the nine levels and their definitions as utilized by the DoD. The TRLs not only provide assistance in acquisition of complex systems but also clearly provides guidelines in the transition from research to development. Although this is not an exact science to date, it is one of the best tools available for the decision makers to determine the level of investment in a program using new technologies. [DoD, 2005]

Table 2-1: Technology Readiness Levels for Assessing Critical Technologies

Technology Readiness Level	Description
1. Basic principles observed and reported.	Lowest level of technology readiness. Scientific research begins to be translated into applied research and development. Examples might include paper studies of a technology's basic properties.
2. Technology concept and/or application formulated.	Invention begins. Once basic principles are observed, practical applications can be invented. Applications are speculative and there may be no proof or detailed analysis to support the assumptions. Examples are limited to analytic studies.
3. Analytical and experimental critical function and/or characteristic proof of concept.	Active research and development is initiated. This includes analytical studies and laboratory studies to physically validate analytical predictions of separate elements of the technology. Examples include components that are not yet integrated or representative.
4. Component and/or breadboard validation in laboratory environment.	Basic technological components are integrated to establish that they will work together. This is relatively "low fidelity" compared to the eventual system. Examples include integration of "ad hoc" hardware in the laboratory.
5. Component and/or breadboard validation in relevant environment.	Fidelity of breadboard technology increases significantly. The basic technological components are integrated with reasonably realistic supporting elements so it can be tested in a simulated environment. Examples include "high fidelity" laboratory integration of components.
6. System/subsystem model or prototype demonstration in a relevant environment.	Representative model or prototype system, which is well beyond that of TRL 5, is tested in a relevant environment. Represents a major step up in a technology's demonstrated readiness. Examples include testing a prototype in a high-fidelity laboratory environment or in simulated operational environment.
7. System prototype demonstration in an operational environment.	Prototype near, or at, planned operational system. Represents a major step up from TRL 6, requiring demonstration of an actual system prototype in an operational environment such as an aircraft, vehicle, or space. Examples include testing the prototype in a test bed aircraft.
8. Actual system completed and qualified through test and demonstration.	Technology has been proven to work in its final form and under expected conditions. In almost all cases, this TRL represents the end of true system development. Examples include developmental test and evaluation of the system in its intended weapon system to determine if it meets design specifications.
9. Actual system proven through successful mission operations.	Actual application of the technology in its final form and under mission conditions, such as those encountered in operational test and evaluation. Examples include using the system under operational mission conditions.

In order to provide a useful overview portrayal of the relationship situation between the level descriptions provided in the TRL table, the Systemigram appearing in Figure 2-8 was created.

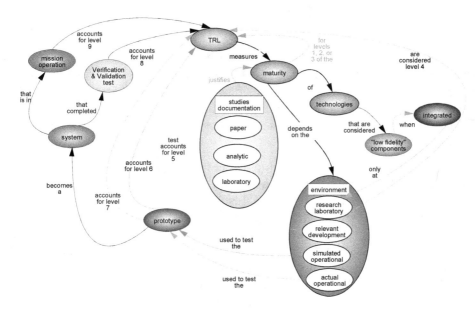

Figure 2-8: Graphical Representation of the TRL Scale

As the reader will observe in examining this example, a Systemigram is a network of nodes and links, flow, inputs and outputs, a beginning and an end that fit onto a single page. Key concepts, noun phrases specifying people, organizations, groups, artifacts and conditions are indicated as nodes. The links between the nodes are verb phrases (sometimes prepositional phrases) indicating transformation, belonging, and being. As illustrated in this situation, some nodes can be collectors for multiple nodes. Nodes can be colored to provide an additional view of relationships between certain nodes. In this figure in black and white only the nuance between the colors are visible. In the TRL example, the reader will observe that the Systemigram expresses relationships that are not as easy to see by simply examining the text in the TRL table.

SystemiTool is an extremely useful tool for creating and animating Systemigrams and it is available from www.boardmansauser.com.

SIMULATING SYSTEM DYNAMICS

The models described in the previous section are qualitative. However, qualitative models do not provide insight in respect to actual dynamics of the situation. While some of the model approaches described earlier have been extended to provide a basis for quantitative system simulation, there are several methods and tools developed explicitly for this purpose. Let us consider one of these approaches to modeling.

As noted above, the first dynamic modeling language DYNAMO was based upon the use of the Fortran programming language and was developed by Professor Jay Forrester at MIT in the 1960s. Inspired by Forester's and Peter Senge's work, Professor Barry Richmond at Dartmouth College led the development of a graphical method for building and simulating systems thinking models. The first modeling tool was STELLA that provided a basis for education and research into the complexities of system situations. STELLA is an acronym for Structural Thinking Experimental Labarotory Learning with Animation. In 1985 STELLA became a commercial product of High Performance Systems (now called isee systems, inc.). To provide a version that focuses upon business systems, in 1990 the company introduced iThink. The main focus is to describe and simulate models in which Stocks (processing places) accept Flows (inputs to be processed). As evaluation proceeds variables reflecting various properties are calculated that can be displayed in table or graph format. Decisions can be made in the execution such as examining thresholds where alternative paths of execution can be made. The reader is referred to www.iseesystems.com for a thorough introduction to these products. The key features of STELLA and iThink as presented by the company are as follows:

- Intuitive icon-based graphical interface simplifies model building
- Stock and Flow diagrams support the common language of Systems Thinking and provide insight into how systems work
- Enhanced stock types enable discrete and continuous processes with support for queues, ovens, and enhanced conveyors
- Causal Loop diagrams present overall causal relationships
- Model equations are automatically generated and made accessible beneath the model layer
- Built-in functions facilitate mathematical, statistical, and logical operations
- Arrays simply represent repeated model structure
- Modules support multi-level, hierarchical model structures that can serve as "building blocks" for model construction

Both STELLA and iThink have been successful products and are widely used in educational, and research in the public and private sectors. Figure 2.9 provides an example of an iThink model.

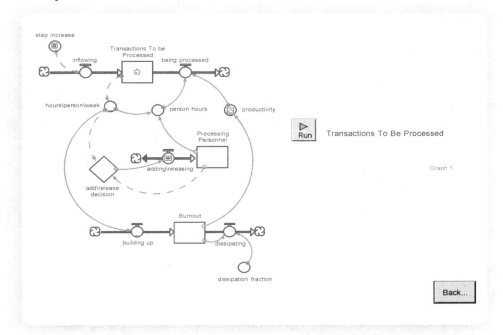

Figure 2-9: iThink Stock and Flow Example (reprinted with permission of isee systems, inc.)

The model shows Stocks as rectangles and Flows from a source and through a valve. The valve can be controlled by external variable (such as the step increase). Variables that are collected include the hours/person/week, person hours, and productivity calculations. The diamond is a decision process that has an effect upon the Flow of adding/releasing personnel. This example is thoroughly described in the free trial download of iThink and the reader is encouraged to take advantage of this offer from isee systems, incorporated.

There are many good examples of applying STELLA and iThink on the isee website. One interesting soft system example related to crises management has been provided by Chris Soderquist [http://blog.iseesystems.com/systems-thinking/we-have-met-an-ally-and-he-is-storytelling]. He exemplifies the usage of STELLA and iThink in exploring the complex relationships of various actors in respect to the anger directed towards the US and Afghan governments.

SOFT SYSTEMS METHODOLOGY

Gwilym Jenkins, the first British Professor of Systems, established in the mid 1960s a center for action research at Lancaster University. Peter Checkland, after fifteen years of experience in scientific, engineering and management roles at ICI Fibers Ltd., joined this effort and has been the leading figure in establishing the basis for what is called "action research" as well the development of a soft system methodology. The research started from a systems engineering perspective where the main idea was to gain knowledge of real-world situations that could then be utilized in improving design and operation of systems that are of interest in an enterprise. As mentioned above, this approach failed to capture situations where objectives and goals could not be clearly established thus enabling the use of engineering methodologies. So the thrust of the efforts for action research and soft systems methodology shifted in focus to dealing with human activity systems.

The aim of the action research is to find ways of understanding and coping with the perplexing difficulties of taking action, both individually and in groups, to improve the situations which day-to-day life continuously creates and continuously changes. In a manner consistent with the Change Management model presented in Figure 1-13, action research involves:

- Undertaking action
- Reflecting upon the insights that may support a deeper appreciation of what is happening
- Insights feed into and improve current actions
- Possible transfer of insight into other domains

These notions of reflections and actions are based upon the establishment of an intellectual framework that defines and expresses what constitutes knowledge of a situation. Checkland indicates that there are linked ideas in the framework, a way of applying these ideas, and an application area. Given the framework, action research has been characterized as follows:

- A collaborative process between researchers and people in the situation
- A process of critical inquiry
- Places focus upon social practice
- Deliberate process of reflective learning

Checkland also points to the need for interpretive based systemic understanding where it is necessary to include the study of the cultural aspects of the situation as well as the interpretations and perceptions of individuals within the cultural context. He therefore points to the need to involve all stakeholders in the action research process.

In order to realize the aims of action research, Checkland successively worked on evolving a methodology for dealing with soft systems composed of complex human activities. The Soft Systems Methodology (SSM) is based upon the need to learn in order to improve some form of *purposeful activity system* which provides some form of T (Transform) of an I (Input) to and O (Output). This system is then captured in the form of a learning system composed of the elements and relationships portrayed in Figure 2-10. The model has been slightly changed in that it is important to consider the pursuit of opportunities as well as problems.

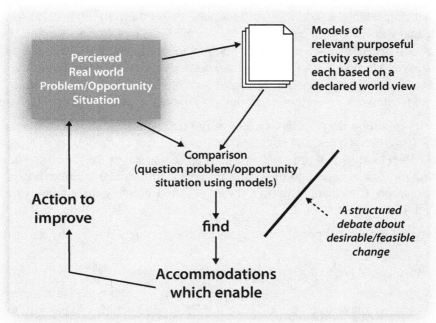

Figure 2-10: Inquiry/Learning Cycle of the Soft Systems Methodology

Associated with the learning system are principles that further characterize the softness of the approach:

- Real world: a complexity of relationships
- Relationships explored via models of purposeful activity based on an explicit world view
- Inquiry structured by questioning perceived situations using the models as a source of questions
- "Action to improve" based on finding accommodations (versions of the situation which conflicting interests can live with)
- Inquiry in principle is never ending; it is best conducted with a wide range of interested parties; give the process away to people in the situation

Note: The methodology is not prescriptive. That is, it does not provide a prescribed sequence of activities leading to a result. In an earlier version of the methodology, Checkland presented a seven-step model that was too often viewed as a prescriptive sequence, even though he pointed to the fact that the methodology could be used at any starting point. The most valuable part of Checklands contribution is that it provides an overall model of the process involved in performing systems thinking studies.

The application of SSM does not provide results that can be proved or disproved as with repeatable scientific investigation. It is only through continual application that the value-added provided by the methodology can be verified. Based upon an earlier seven-step model of the methodology and the current version as portrayed in Figure 2-10, Checkland and his colleagues and students have applied the methodology hundreds of times and are convinced that it has proven to be a useful framework for studying and building the basis for improving soft systems.

In gaining a deep understanding of the real world problem as a purposeful activity system, Checkland suggests the development of models as rich pictures (described earlier) that are cartoon like representations of the situation showing the major stakeholders and the issues involved. In formulating an expression of the situation, Checkland proposed the use of the following three analyses of the situation:

Analysis One – list of possible, potential problem owners selected by the problem solver.

Analysis Two – based upon a roles/norms/values framework

Analysis Three – builds on what any social grouping quickly acquires; a sense of what you have to do to influence people, to cause things to happen, to stop possible courses of action, to significantly affect the actions the group or members take. Views of what is required to be powerful can include knowledge, a particular role, skills, charisma, experience, commitment, etc.

In modeling purposeful activity in order to explore real-world action based upon the Analysis One, Two and Three, it becomes obvious that many interpretations of a declared purpose are possible. Thus, before modeling choices of a relevant world-view of the situation have to be made and declared. Checkland has indicated these choices in the form of a CATWOE mnemonic that indicates the following:

C – Customers who order the study

A – The Actors involved

T – The purposeful activity system Transform to be performed

W – World View of the situation to be studied

O – Owners of the systems involved

E – Environmental constraints

By developing a handful of models, as indicated in Figure 2-10, based upon pure purposeful activity as opposed to descriptions of the real world, a basis for determining the appropriate questions to ask concerning the real situation evolves. Checkland provides suggestions as to model building based upon activities. However, it is here where models based upon structured text; such as the Five Whys, the Loops, Links, and Delay language, Influence Diagrams, Rich Pictures and Systemigrams or dynamic simulation solutions can also be applied in gaining insight.

Based upon the models a structured debate as to desirable change can be made. The next part of the model is to find accommodations for making improvements. Finally, the model indicates that action is taken to improve the system. After this, in a typical feedback fashion, new problem situations or opportunities can arise that require adjustment of the system or the consideration of a new system. While Checkland does provide a link to action to be taken, his methodology does not provide any structured means of accommodating change. Structural changes of institutionalized systems are most often related to systems life cycle management according to system life cycle models as described in the following chapter on acting in terms of systems. A more recent book by Checkland and Poulter provides a short definitive account of Checklands Soft System Methodology and is recommended as a starting point for those interested in pursuing this work [Checkland and Poulter, 2006].

PRINCIPLES TO REMEMBER

Regardless of the models, methods and tools used in supporting systems thinking, there are some general principles that provide some "food for thought" concerning systems thinking as recommended by [Senge 1990]:

- Today's problems come from yesterday's "solutions".
- The harder you push, the harder the system pushes back.
- Behavior grows better before it grows worse.
- The easy way out usually leads back in.
- The cure can be worse than the disease.
- Faster is slower.
- Cause and effect are not closely related in time and space.
- Small changes can produce big results but the areas of highest leverage are often the least obvious.
- You can have your cake and eat it too but not at once.
- Dividing an elephant in half does not produce two small elephants.
- There is no blame.

The more you learn and understand, the more you understand how much more there is to be learned.

KNOWLEDGE VERIFICATION

1. Identify some hard systems and some soft systems as well as examples of mixed systems.

2. What are the dangers associated with relying upon reductionism and simple causal relationships?

3. Identify some examples of paradoxical situations.

4. Beginning with a problem with which you are familiar, work outward towards deeper understanding via the Five Why's approach.

5. Explain a system situation that you are familiar with in terms of Influence Diagrams.

6. Explain a system situation that you are familiar with in terms of Links, Loops and Delays.

7. Utilizing reinforcing and balancing loops describe a systems related situation in your organization involving growth and limiting factors.

8. What is the purpose of a system archetype?

9. Create a rich picture of a complex situation with which you are familiar; perhaps related to your work or free time activities.

10. Given a prose description of a problem or opportunity system situation, create a Systemigram.

11. Augment the prose description and the Systemigram from 10 with a respondent system aimed at handling the situation.

12. What is meant by action research and how does the soft systems methodology relate accomplishing the aims of the research?

13. Using a tools such as STELLA or iThink develop a model of processing situation where materials or service flows are being processed.

14. How can concepts of systems thinking as provided in this chapter be utilized in your organization (enterprise)?

Interlude 1: Case Study in Crises Management

This case study is based upon a project with the title: "A Systems Look at Special Response Units in Law Enforcement" done by Rich Wright as a part of a graduate course presented by your author and provided by the Stevens Institute of Technology in Baltimore during the spring of 2009.

Abstract

This project will use a systems-oriented methodology to study the history and evolution of Special Response Units (sometimes better known as Special Weapons and Tactics (SWAT) Teams) by law enforcement organizations since the 1960s. SRUs can be seen as respondent systems used to counter situations (e.g. hostage takings) that traditional police forces were poorly equipped to deal with. The strengths of SRU systems will be examined, in addition to problems with their use and possible solutions.

INTRODUCTION

We Have a Situation...

In August 1965, the Watts neighborhood in Los Angeles, California erupted in race riots. The 1972 Olympic Games in Munich Germany became the scene of hostage taking and murder by terrorists. In 1992 and 1993 extremists and federal agents were killed during shootouts and sieges at Ruby Ridge, Idaho and Waco, Texas. At a high school in Columbine, Colorado, two students shot other students and faculty in 1999. While separated by years of time and thousands of miles, these events had more in common than just history-making violence. They all functioned as situation systems that provoked dramatic changes in law enforcement response systems, both in the United States in other parts of the world. [Cullen, 2009],[Gates and Shah, 1992],[Halberstatd, 1993],[Lloyd, 2009],[Walmer, 1986] and [Whitcomb, 2001]

What is an Special Response Unit?

A Special Response Unit is a generic term for any law enforcement unit charged with responding to situations that standard law enforcement units lack the training, equipment or other resources to handle. The traditional set of those situations, for which the first SWAT (Special Weapons And Tactics) team was created, includes snipers, barricaded suspects, hostage rescue, and high-risk warrants (those involving suspects known to be well-armed or having a record of violent behavior). Over time some SRUs have had their missions and capabilities expanded so they can also raid drug labs, protect Very Important Persons (VIPs), perform counterterrorism operations and more. [Federal Bureau of Investigation, 2009]

SRUs are often formed as part of a city, county, state, or regional (e.g. several cities or counties) police force. They can also be federal units. Some federal SRUs have a limited Area Of Responsibility such as military police Special Reaction Teams (SRTs) that protect American military installations [Department of the Army, 1987]. Others like the FBI's Hostage Rescue Team [Federal Bureau of Investigation, 2009] or Germany's GSG 9 (Grenzschutzgruppe 9) [Walmer, 1986] have national and international jurisdiction to protect their nation's citizens and interests.

BACKGROUND

Los Angeles and the Birth of SWAT

The riots in Watts in 1965 would go on record as the worst of the riots of the 1960s, with 34 dead, 1032 injured, over 600 buildings burned or looted and 3438 arrests. [Gates and Shah, 1992] Less well known than those statistics are the hundreds of sniper attacks attempted against police during the riots. Inspector Daryl Gates, later Chief of the Los Angeles Police Department (LAPD), recalled, "We had no organized response to snipers, so the police would shoot back indiscriminately. By the time I would arrive, everybody was blasting away. It was not easy to get them to stop." [Gates and Shah, 1992] Gates saw the same problem again only a month later when a man barricaded himself in his home and began shooting. The suspect held off large numbers of officers, wounding three of them and a civilian before he was wounded and arrested. [Halberstatd, 1993]

To deal with snipers and barricaded suspects, Gates and several of his colleagues began studying guerilla warfare and the counterinsurgency tactics then being used by the U.S. military in Vietnam. They worked with Marines at Camp Pendleton and the Chavez Ravine Naval Armory to help slowly train some of the department's best marksmen as counter-snipers. By 1967, Gates was in charge of LAPD's Metro Division. He reorganized Metro into military-style platoons and was finally able to consolidate his 60 marksmen into Platoon D of that division. Gates proposed a name for the new unit, Special Weapons Attack Teams (SWAT) to Deputy Chief Ed Davis. Davis objected to the military sound of the name, so Gates adjusted it to the now famous "Special Weapons And Tactics". The Deputy Chief approved immediately. [Gates and Shah, 1992]

In spite of SWAT's law enforcement mandate to get all victims, suspects and officers out of a crisis alive, the military tactics and weapons they adapted made them outcasts among the rest of LAPD. Their weapons and equipment had to be purchased by the SWAT members themselves, salvaged, or bought from surplus stores. [Halberstatd, 1993] SWAT teams trained at Camp Pendleton with the Marines, on the back lots of Universal Studios and other out-of-the-way locations. Gates' autobiography indicates that although he wasn't responsible for SWAT tactics, he did offer whatever resources and encouragement he could. This included role-playing as the 'hostage' while SWAT teams slowly added hostage rescue to their list of competencies. [Gates and Shah, 1992]

SWAT's first official mission was serving search and arrest warrants against 2 members of the Black Panthers organization on 8 December 1969. The Panthers were effectively a nation-wide gang with political ambitions to lead the "peoples' revolution". They had assaulted several LAPD officers with deadly force. LAPD intelligence indicated that the Panthers building was heavily fortified with sandbags, gun portals, escape tunnels and loads of weapons. SWAT was the best-equipped unit to serve the warrants. However, in spite of the intelligence and a week of planning, the initial assault resulted in three SWAT members wounded and no Panthers in custody. With the element of surprise lost, the fortifications made another direct assault impossible. A standoff developed with SWAT and the Panthers trading automatic weapons fire for hours, forcing an evacuation of the surrounding area. SWAT requested a grenade launcher from the Marines at Camp Pendleton. The request went through Gates (who was Acting Chief while LAPD's Chief was in Mexico), to the Mayor's office all the way to the Pentagon, where it was approved. After the grenade launcher quietly arrived, Gates directed SWAT to make one more plea for the Panthers to surrender, before unspecified "drastic measures" were taken. Fortunately, the six Panthers surrendered under a makeshift white flag and were arrested. The SWAT blunted assault and the shootout became lessons learned due to the lucky success with no loss of life. The presence of the grenade launcher, which could have transformed LA into an urban war zone only three years after Watts, was unknown to the media and the public for many years. [Gates and Shah, 1992]

The Black Panther arrests secured the future of SWAT with the LAPD and led other law enforcement organizations at the local, state and federal level to create their own SWAT teams. The U.S. Marshals created one of the first federal SWAT teams, known as their Special Operations Group (SOG) in 1971. [U.S. Marshals Service, 2009] The FBI followed suit [Halberstatd, 1993], eventually establishing SWAT teams at each of its 56 field offices to assist field agents and local law enforcement.

FBI and LAPD SWAT teams met in 1974 on the trail of Simbionese Liberation Army (SLA) members wanted for murder, bank robbery and kidnapping and brainwashing newspaper heiress Patty Hearst. Like the Black Panthers, the SLA was heavily armed. In an operation reported on national TV, SWAT surrounded the house and made surrender demands that were ignored. Their tear gas grenades were met with gunfire from a Browning Automatic Rifle (BAR) and other guns. As with the Black Panthers, SWAT requested military-grade weaponry. This time it was fragmentation grenades. Not wanting to cross into military warfare Gates refused. [Gates and Shah, 1992] The siege and shootout had continued for 50 minutes when the SLA house caught fire. However, the SLA kept shooting and firemen refused to approach. The house burned to the ground killing everyone inside while law enforcement officers watched helplessly. While the outcome was less favorable than that of the Black Panthers incident, The SWAT units performed their job and became recognized throughout the law enforcement community and around the world. [Halberstatd, 1993]

Munich and the Shadows of Terror

Between the LAPD's operations against the Black Panthers in 1968 and SLA in 1974, the world changed forever in 1972. Palestinian terrorists known as Black September infiltrated the Olympic village in Munich Germany. They killed two Israeli athletes and took nine hostages. Attempts by German security authorities to resolve crisis ended in a shootout that killed all the hostages and terrorists. With fears of appearing militaristic, the Germans had opted for 'low key' security at the Olympics and had not developed a dedicated unit to combat the terrorist threat that had begun blooming in the 1960s. Munich underscored these mistakes and the Germans responded by creating a dedicated anti-terror force. However, unlike the units established by other nations' militaries, the German unit was part of their Federal Border Police. This gave it authority to operate within Germany during peacetime and minimized associations with former elite Nazi military units like the notorious SS. [Walmer, 1986].

The new unit was named the Border Guard Group 9 (Grenzschutzgruppe 9 or GSG 9), because the Border Guard already had eight other units. GSG 9 was trained with Israeli help and its leader accompanied Israeli military commandos on their well-known raid on Entebbe Airport in Uganda in 1976, which freed hostages from a hijacked Air France flight. GSG 9 secured its own fame by liberating hostages from a Lufthansa airliner that was hijacked to Mogadishu, Somalia in 1977 by the terrorist Red Army Faction. [Walmer, 1986].

The events of Munich and the successes of the Israelis and GSG 9 left an impression on the US. In 1977, the U.S. Army authorized creation of its anti-terrorist unit known as the Delta Force. The seizures of the US embassy in Iran in 1979 brought the threat even closer to home. In 1980 came Delta Force's failure to rescue the Iranian hostages and the success of Britain's Special Air Service (SAS) to liberate the Iranian embassy in London from terrorists. [Walmer, 1986]

When the US was chosen to host the 1984 Olympic Games in Los Angeles, the FBI and others foresaw the possibility of another Munich on US soil. They used this as the impetus to create a dedicated national counter-terrorism unit called the Hostage Rescue Team (HRT), similar to GSG 9 [Gates and Shah, 1992]. HRT was designed as a step above the SWAT teams the FBI already had at its field offices. It is a self-sufficient 50-man unit designed to deploy from Quantico to anywhere in the US in a matter of hours. HRT flies in its own men, weapons, trucks and even helicopters in a USAF Reserve cargo plane. It can call on the resources of the entire bureau for intelligence and other resources, but its members have no collateral duties of typical law enforcement such as investigations. They serve as a national asset to carry out large-scale anti-terrorism, hostage rescue and special operations that most SWAT units aren't capable of. [Whitcomb, 2001]

SPECIAL RESPONSE UNIT CASES OF INTEREST

The following two cases are described in further detail since later in the project report factors relevant to these SRU's will be highlighted.

Ruby Ridge and Waco

On 21 August 1992, a member of the U.S. Marshals Special Operations Group was killed in a gunfight with the family of Randy Weaver at Ruby Ridge in Idaho. SOG had been serving a warrant. Weaver's son was also killed. Weaver was a former Green Beret and survivalist, and known to be heavily armed. The Marshals retreated and requested assistance, which arrived in the form of the FBI HRT on 22 August. The FBI assumed responsibility and HRT snipers secured a perimeter around the Weaver cabin. Acting under modified rules of engagement, one of HRT's snipers wounded Randy Weaver and killed his wife. The local and state police and National Guard were all mobilized. Negotiations dragged on for ten days before the survivors surrendered. Ruby Ridge became a public relations flashpoint for the FBI and the facts of the shootings were the subject of FBI, Department of Justice and Congressional inquiries over the next several years. [Whitcomb, 2001]

Only a few months after Ruby Ridge, HRT inherited another failed operation from another agency. On 28 February 1993, four agents of the Bureau of Alcohol, Tobacco and Firearms (ATF) had been killed and sixteen others wounded in a shootout. They had been serving a search warrant on the compound of the Branch Davidian religious group in Waco, Texas. Before the raid, the ATF tipped off a reporter, who inadvertently tipped off the Branch Davidians. ATF proceeded with the raid, even after warnings by an informant at the compound that it was expected. [Halberstadt, 1993] As at Ruby Ridge, HRT surrounded the compound with support from local law enforcement and the National Guard. The Davidians were heavily armed and supplied with food, expecting an apocalyptic showdown. Their weapons included pistols, semi-automatic rifles and .50 caliber machine guns. [Whitcomb, 2001] After a 51-day siege, an assault was begun on the compound. Armored vehicles made holes in the walls and pumped in tear gas. Armored personnel carriers carrying HRT assault teams stood by. HRT's rules of engagement stated that they could only return fire, and only if they visually identified specific human targets that posed specific threat. Later, fires began and the compound burned to the ground. Nine Davidians escaped the flames, while 75 others perished. However, HRT fired no shots in the assault. [Whitcomb, 2001] While the details and legalities of Ruby Ridge and Waco have been thoroughly examined and debated in other forums, these tragedies demonstrate several Command and Control (C2) lessons worth exploring for national-level SRUs.

Columbine

While the events of Ruby Ridge and Waco made some in law enforcement more hesitant to 'pull the trigger', those on the other side of the law weren't. On 20 April 1999, two students at Columbine High School near Denver, Colorado went on a shooting rampage. They killed 13 students and wounded 21 others within just 16 minutes, and then committed suicide. [Lloyd, 2009] SWAT Teams didn't arrive in time to intervene and were hampered by alarms they couldn't shut off and inaccurate floor plans. [Cullen, 2009]

The lessons from Columbine were many. Children can ruthlessly gun down other children. Some shooters don't negotiate and aren't afraid to die. SWAT Teams need up to date plans for assaulting schools. And time spent waiting for SWAT can sometimes be measured in lives. Many of these were not new lessons in law enforcement, but Columbine refreshed them and brought them to general public awareness by national television. Security measures at schools were increased, and police departments rethought their approaches to this scenario, which could happen anywhere there is a school. One of these approaches, called "Active Shooter Protocol"[Cullen, 2009] is worth examining for its shift away from SRU reliance and possible long-term consequences.

ANALYSIS OF SPECIAL RESPONSE UNITS AS SYSTEMS

Classification and Topologies of SRU Systems

In order to begin analyzing SRUs as systems, it must be determined what type of system they are. According to the definitions of system types provided in Chapter 1 of this book, several possible classifications, which include natural systems, defined physical systems, defined abstract systems and human activity systems. SRUs are best classified as human activity systems, since they are made up of humans and, per the classification criteria, "are consciously ordered in wholes as a result of some purpose or mission."

The mission of an SRU is to respond to crisis situations as previously discussed. The requisite 'conscious order' can be confirmed by examining an SRU's topology. A system can have a hierarchical topology, a network topology, or both at various points in time. A hierarchical topology results from "a defined system that is developed to meet some need. In this case, the need is synonymous with an SRU's mission, and its hierarchical topology can be seen in a sample SRU organizational chart in Figure 1.

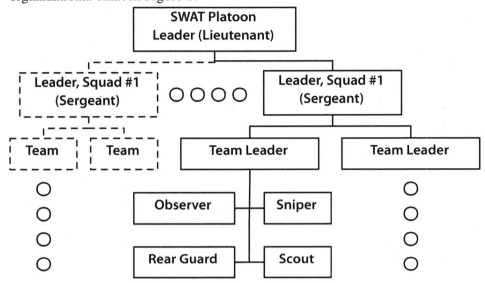

Figure 1: Organization of LAPD's SWAT Unit in 1967

While an SRU's chain of command is always a clearly defined hierarchy, a network topology emerges when it deploys. This network is made up of various quantities and sizes of teams, which are established depending on the SRU's size and the situation. Such an operational topology is best described in a structured text and then visualized with a Systemigram, like that shown in Figure 2. [Boardman and Sauser, 2008].

The structured text is: "For deployments, an SRU is typically organized into one or more Assault Teams, one or more Sniper Teams and sometimes one or more Medics. The SRU Commander in the Tactical Command Post (TCP) commands the Assault Team, Sniper Team, and Medics. A Sniper Team monitors the target, reports target information to the TCP, protects an Assault Team and may shoot the target. Sniper Teams are made up of a Sniper and an Observer who protects the Sniper. An Assault Team assaults the target and reports status to the TCP. Within an Assault Team, a Rear Guard protects Arrest Teams, who follow Pick-up Men, who protect Cover Men, who cover Point Men, Medics aid Assault Teams and Sniper Teams." [Halberstatd,1993]

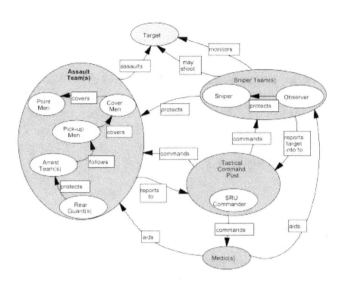

Figure 2: Network Topology of a Deployed SRU

Special Response Units as Respondent Systems

Knowing an SRU's system classification and topologies only provides a partial understanding of it. How is an SRU "system" instantiated for deployment? What enables an SRU to morph its topology to an optimal configuration? The answers to these questions require a larger context, which can be gained from a system-coupling diagram like the one shown in Figure 3.

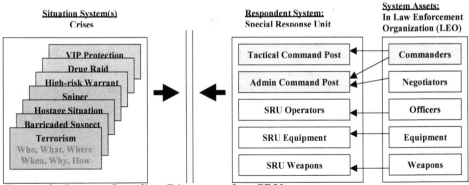

Figure 3: System-Coupling Diagram of an SRU

An SRU is as shown in the center of Figure 3, and constitutes our Narrow System of Interest (NSOI). It is a respondent system, which means it is formed in response to some situation. The various types of situations that an SRU responds to are shown on the left side of Figure 3 and are familiar from the list of SRU "missions" discussed earlier. Each situation can also be viewed as a situational system that arises from a set of circumstances described by the 5 W's (who, what, where, when, why) and a "how" as shown in the figure. On the right side of the diagram we see that an SRU is instantiated from a set of assets that belong to its parent Law Enforcement Organization (LEO). The LEO is also a system of human activity, with a broader mission to enforce law. The situation system and LEO system, when considered together with the SRU, make up the Wider System of Interest (WSOI).

Critical Factors in SRU System Adaptability

Since the assets of a LEO are finite and therefore the assets of an SRU even more so, the key to any SRU's success is its adaptability. Three critical factors in SRU adaptability are intelligence, experience and command elements.

An SRU gains intelligence about a situation in different ways, depending on whether that situation is deliberate or an emergency. A deliberate situation is one that the SRU knows about in advance, either because the SRU sets the timetable (e.g. for a high risk warrant or drug raid) or receives advanced notice (e.g of a VIP's arrival). Typical emergency situations, which occur without warning, are sniper attacks, hostage situations and barricaded suspect or terrorist action. In a deliberate situation, intelligence is gathered in advance and provided to the SRU in an operation plan or warrant. In an emergency situation, a 911 operator or police dispatcher gathers initial intelligence by asking questions listed in a Standard Operating Procedure for that emergency type, and then more is gathered by the SRU at the scene.[Halberstatd, 1993]

SRUs gain experience by responding to both real situations and thematic ones. Thematic situations take the form of training scenarios developed for and by the SRU. They can be based on prior real situations, or hypothetical ones. Training is beneficial because it allows situations that are normally encountered only as emergencies to be experienced in a deliberate environment where they can be repeated and manipulated as necessary for learning without unnecessary risk.

The third critical factor in SRU adaptation is the presence of control elements. This includes Tactical Commanders (SRU leaders) and Administrative Commanders of the LEO (e.g. a Watch Commander in a large LEO or police chief in a small one). Tactical Commanders are fluent in SRU assets and use both available intelligence and their experience to configure the unit to meet the situation. This includes determining the need for assault, sniper and medical teams, team size and makeup, equipment and even individual team assignments. Tactical commanders operate in a TCP close to the situation and control the SRU and keep Administrative Commanders informed. Administrative commanders usually operate at a more distant Administrative Command Post, coordinating support from other LEO assets (e.g. hostage negotiators, intelligence or patrol officers) or external services (e.g. fire, medical) while communicating with higher (often political) authorities. [Halberstatd, 1993].

SRU System Problems and Solutions in the Cases of Interest

Command and Control

While command elements are one of the factors that govern the success of SRUs, problems with Command and Control (C2) can also be a source of SRU weakness. This is especially for national teams with more complex control chains like HRT and GSG 9. Some of these problems include command handoffs, administrative control, burden of control and response control.

In the majority of emergency situations, the standard LEO units that arrive first assume command. When the determination is made that they cannot handle the crisis, the most convenient SRU is contacted. When the SRU arrives, there is a handoff of full operational control is from the initial LEO commander to an SRU commander. For many SRU operations, such handoffs occur between Patrol and SWAT leaders within the same department (e.g. LAPD to LAPD SWAT) who often work together. [Cullen, 2009] However, national teams like HRT have an increased likelihood of handoff from a completely separate agency, and this may occur after a prior handoff between LEO and SRU units within the first agency (e.g. U.S. Marshals to U.S. Marshals SOG). Knowledge is always lost in such a transfer, and the greater the operational 'gap' of unfamiliarity between the parties, the more risk to the operation. In the era of Ruby Ridge and Waco, national-level SRUs like FBI's HRT and US Marshal's SOG were designed to operate independently, so as to reduce dependencies for deployment and minimize impact on local LEOs. However, this approach also decreased their likelihood of their training together and meant they were less familiar with each other's operations. Command handoffs between independent national agencies occurred at both Ruby Ridge and Waco. [Halberstatd, 1994] and [Whitcomb, 2001] .

The next C2 problem involves administrative control. While tactical command at crises belongs to an experienced SRU team leader, administrative command differs between national and non-national teams. For city, county, state and even FBI SWAT teams, their administrative command is typically a police chief or agent with direct law enforcement experience. For national-level teams like HRT and GSG

9, however, administrative management often includes political leaders (like the Attorney General in HRT's case). These leaders are typically far removed from the crisis site, which can induce decision delays. Also, in some countries like the U.S. many political leaders have no police or military experience to aid their decisions. HRT experienced both delays and indecision from their national command level during both Ruby Ridge and Waco. [Whitcomb, 2001].

The third C2 problem involves the burden of control. While national teams may be independent in terms of their own equipment, they cannot sustain themselves during extended operations that run weeks or months. Continuing such operations requires more assets (typically local) of different types. This increase of assets and their dissimilarity combined with increased media, budgetary or political scrutiny dramatically increases the burden of control placed on those in command of national SRUs. This can in turn increase pressure to reach a resolution quickly. This increased burden of control was present at both Ruby Ridge and Waco. [Whitcomb, 2001] .

The last C2 problem involves controlling a response. While there are benefits to a highly capable SRU assisting other agencies, its competencies must be considered in the context of the situation. At both Ruby Ridge and Waco, a Hostage Rescue Team optimized to deal with terrorism incidents was employed against groups that were not necessarily terrorists, in situations where it was unclear that anyone was being held hostage. [Whitcomb, 2001] Furthermore, did arresting a family of six in the mountains of Idaho require a national asset, or just one more experienced with wilderness operations than those initially employed? Does the fact that a group has lots of weapons require a national response unit, or just lots of well-supported local ones? Is an initially disastrous assault better followed up with bigger, better team or just a better plan? Given the speed with which HRT was summoned to Ruby Ridge and Waco, it is unclear how much consideration leaders in the Marshals, ATF or FBI gave to such questions.

In the aftermath of Ruby Ridge and Waco, the next FBI director, Louis Freeh, worked hard to address these C2 problems. One of the main ways he did this was to create the Critical Incident Response Group (CIRG), which integrated HRT with many other FBI units that play roles in response to crises nation-wide, such as hostage negotiators, behavior specialists, and logistics personnel. [Federal Bureau of Investigation, 2009] and [Whitcomb, 2001] CIRG's integration helped slowly break what had been an almost isolationist mindset of independence for units like HRT. First, this integration helped reduce the "gap" when command handoffs occurred within the FBI and potentially between FBI and other agencies. Secondly, GIRG's leaders were given nation-wide authority, which presumably gave it more decision making authority in crises and reduced issues of administrative control with Washington leaders. Thirdly, the integration of logistics personnel and liaisons within GIRG reduced the burden of control during extended operations. This was exemplified by FBI's ability to successfully manage an 81-day siege with the Montana Freemen group in 1996 that ended in their surrender without loss of life. [Whitcomb, 2001] Lastly, with CIRG, the FBI rethought its responses to such confrontations. Negotiation became the watchword, and as demonstrated with the Freemen, time was not an issue. HRT members were even cross-trained in negotiation. [Whitcomb, 2001].

Response Time

One of the potential weaknesses of a respondent system is the time it takes for that response. Most LEOs lack the staff or funds to maintain a dedicated SRU at constant readiness like HRT, so the members often have "day jobs" unrelated to SRU, such as being patrol officers or detectives, and their SRU role is a secondary duty performed as needed. This dispersal of assets within a LEO has a negative effect on deployment time, which has been recognized since the LAPD formed the first SWAT units. Both SRU personnel and their equipment must reach the crisis to be effective. If the equipment is all stored at 'headquarters', the time to deploy can become considerable. To counter this, some SRUs allow their members to carry their gear and weapons with them during their regular duties. [Gates and Shah, 1992] and [Halberstatd, 1994] Other LEOs store SRU gear in a central location, but transport it to the scene in a large vehicle while the team is en route and the team equips itself at the TCP. [Halberstatd, 1994]

Even when these approaches are taken, however, the time for travel and suiting up with full body armor and heavy weapons is unavoidable. This fact was brought home by the Columbine tragedy. Since Columbine, some LEOs are providing their officers with training, equipment, and authority to intervene in "active shooter" situations without waiting for SRUs or even standard backup units. This is known as Active Shooter Protocol. The equipment varies, but includes at least a ballistic shield that provides some minimal amount of protection and can be used with the officer's own sidearm. Other departments go further and provide semi-automatic rifles to increase firepower. Patrol units are trained to use this equipment and to stop an active shooter as soon as possible, ignoring even the wounded victims. They are also trained as to the limits of this authority and their equipment. Risks in taking such independent action are high for all involved, but the possibility of saving more lives encourages this alternative, and the debate. [Cullen, 2009] and [Lloyd, 2009]

Escalation

One of the continuing threats to the effectiveness of SRUs and LEOs in general involves escalation. It is described in one of the final scenes of the superhero movie "Batman Begins" [Nolan, 2005], during a discussion between Batman and Police Lieutenant Gordon about how to restore order to the now chaotic city of Gotham

Batman: "...We can bring Gotham back."   Lieutenant Gordon: "What about escalation?"

Batman: "Escalation?"

Lieutenant Gordon: "We start carrying semi-automatics, they buy automatics. We start wearing Kevlar, they buy armor-piercing rounds."

While Batman could be viewed as a one-man SRU for the Gotham Police Force, Gordon's statements reflect concerns about adopting Batman's equipment and tactics too broadly. While potentially effective in the short term, this approach can trigger an arms race with criminals in the long term. When this phenomenon was considered using the data on criminal situation systems and LEO/SRU systems gathered for this paper and described using Senge's graphical language of links and loops [Senge, 1990], Figure 4 emerges.

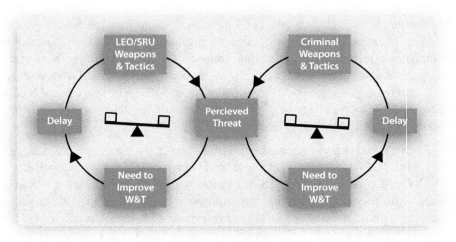

Figure 4: Balancing Loops showing Escalation between LEO/SRU and Criminal Weapons and Tactics

Key points to note in the figure are that the perceived threat is what drives the loops, and it may or may not be real at all times. The delays are also critical, because they are what make the escalation trend harder to detect and can give either side a false and temporary sense that the threat has been "beaten". Delays can be caused by shortages of funding, technology or knowledge needed to improve weapons and tactics.

In terms of real world applications, the birth of SRUs with LAPD back in the 1960s can be traced to LAPD's correct perception that criminals were using new weapons (e.g semiautomatic rifles) and tactics (e.g. snipers and barricades). [Gates and Shah, 1992],[Halberstatd, 1994] While there was not an immediate or universal response from criminals to the development of SRUs (indicating a delay), gradual improvements have been seen among criminals with sufficient funds and organization, such as drug dealers. For instance, LAPD SWAT's traditional entry methods became less effective for assaulting drug labs when the dealers began using bulletproof steel doors, security systems and heavy weapons to defend their labs. To regain the upper hand, SWAT acquired "Armored Rescue Vehicles" to resist criminal gunfire and affixed battering rams to them to knock down the steel doors. [Gates and Shah, 1992] Similarly, groups wishing to resist LEOs and SRUs have sometimes armed themselves more heavily. This can be seen in the reports of .50 caliber machine guns in possession of the Branch Davidians at Waco, which led to HRT's request for National Guard armored personnel carriers. [Whitcomb, 2001] In the future, if LEOs choose to propagate the weapons and tactics of "Active Shooter Protocol" to additional scenarios "to protect the public", criminals may begin perceiving every LEO officer as a one-man SRU and improve their weapons and tactics in response.

The only way to minimize escalation is for one side to minimize their responses whenever risks permit. This responsibility will fall to the LEOs and SRUs, since it is unlikely for criminals to do so. This response minimization will reduce the threat perceived by criminals, and lower the risk to everyone. The rewards in this approach have been demonstrated by the FBI's current preference for negotiations before weaponry, which has so far prevented repeats of Ruby Ridge and Waco.

CONCLUSIONS

This project has reviewed the history that led to the development and evolution of SRUs and examined them both as Narrow Systems of Interest and within the context of Wider Systems of Interest. SRUs are human activity systems and as their name implies, they are responsive systems. Their strength lies in their ability adaptability, which is achieved by command elements configuring LEO assets to meet situations, using intelligence and experience. Using cases of interest, SRU systems problems with command and control, response time and escalation were examined. Systems solutions were also examined, though none of these are permanent due to the ever-changing environments that SRUs and LEOs face. Areas for possible future research include using the links and loops technique to examine SRU interactions with LEOs that may demonstrate archetype scenarios such as "accidental adversaries", "shifting the burden to interveners", and "tragedy of the commons".

CASE STUDY REFERENCES

Boardman, J. and Sauser, B. (2008) Systems Thinking: Coping with 21st Century Problems, CRC Press, Boca Raton, FL.

Cullen, D. (2009) The Four Most Important Lessons of Columbine, Slate Magazine, 29 April 2009. Retrieved from http://www.slate.com/id/2216122/ May.

Department of the Army (1987) Field Manual 19-10: Military Police Law and Order Operations

Federal Bureau of Investigation (2009) FBI Tactical Hostage Rescue Team. Retrieved from http://www.fbijobs.gov/116.asp

Gates, Daryl F and Shah, Diane K. (1992) Chief: My Life in the LAPD, Bantam Books.

Halberstatd, Hans (1994) SWAT Team: Police Special Weapons and Tactics.

Lloyd, Jillian (2009) Change in tactics: Police trade talk for rapid response, Christian Science Monitor, 31 May 2009. Retrieved from http://www.csmonitor.com/2000/0531/p2s2.html

Nolan, Christopher (2005) Batman Begins. New York: Time Warner.

Senge, Peter M. (1990) The Fifth Discipline: The Art & Practice of The Learning Organization, Currency Doubleday, New York.

U.S. Marshals Service (2009) Historical Perspective. Retrieved from http://www.usmarshals.gov/history.index.htm

Walmer, Max (1986) An Illustrated Guide to Modern Elite Forces, Salamander Books Ltd.

Whitcomb, Christopher (2001) Cold Zero: Inside the FBI Hostage Rescue Team.

Interlude 2: Case Study in Organizational Development

This project entitled "The Company Culture of Handelsbanken as a System" was provided by Susanna Göransson as a part of a course presented by your author at the Mälardalen University in Västerås, Sweden during the fall of 2008.

Abstract

Handelsbanken is a decentralized organization that has been very competitive for many years. Instead of controlling the organization by detail, Handelsbanken policy indicates that decisions are to be made at the lowest organizational level that is possible. This means that each branch office has a large amount of freedom to do what they think is the best, but at the same time following the spirit and principles of Handelsbanken. "Mål och medel" is a set of principles and beliefs that everyone in the bank should have as a guideline when making decisions in their everyday work.

The intention of this project is to apply systems thinking in understanding the company culture of Handelsbanken. In which ways does the bank describe its company culture? Can the values and beliefs of the bank which are supposed to lead to higher financial performance as stated in "Mål och Medel" be described as a part of a system that also includes financial elements? Some different systems thinking tools have been used in order to understand the company culture as a system. A "Rich Picture" has been drawn an used as a first tool in order to get an overall picture of the issues involved, a "Systemigram" has then been drawn that in a systemic way describes what "Mål och Medel" says in writing and finally the company culture has been analyzed as a Systems of Interest.

INTRODUCTION

Handelsbanken is an old Swedish bank which has been highly competitive on the national and international market for many decades and today has 10,500 employees. It has 660 branch offices in 21 countries, making it the most international bank in the Nordic countries[1]. Handelsbanken's share was listed on the Stockholm stock exchange already in 1873, and is therefore the oldest of the shares currently listed on the exchange. So there is an old heritage of banking in the company.

A DECENTRALIZED ORGANIZATION

A large re-organization was carried out in the early 1970's by the then president Jan Wallander when the company headquarters was dramatically reduced in size and many functions were decentralised to regions or even down to the office level. The lines of reasoning behind this move were based on a humanistic view of man as being proactive and meaning making. The employees were seen as capable and motivated in terms of utilizing the increased scope for action provided by the decentralization, which enhanced their possibilities of making decisions at local branch level. Core characteristics introduced by Wallander [Wallander, 2002] were:

- A "humanistic" view of man as (potentially) active, responsible and development oriented. The point of departure was based on Maslow's hierarchical theory. Consequently, the operative level of the organization consists of relatively small units (local branches) with quite a high level of decision-making possibilities
- No central budgeting since the 1970's, which made the bank differ substantially from other banks and companies in Sweden at that point in time. Decisions about credit, employment, work tasks, promotion and salaries took place at local level
- Generous pension funds for all employees including profit-sharing, which has been quite substantial due to the high competitiveness of the company over the past decades
- A regulation system for improving competitiveness. All local units have access to result and balance sheet information and are thus aware of if and how they and other units and banks contributed to profitability. The aim is to achieve above-average cost-income quota levels compared with other branches, which is intended to generate continuous pressure and ensure that each branch remains competitive

1 . www.handelsbanken.se

- Management by objectives. As a consequence of the humanistic view, the employees are considered sufficiently competent to regulate their work on a local level. The aim is the formulation of clear and meaningful goals, which make it possible to evaluate how individual performance contributes to goal fulfilment

HANDELSBANKEN TODAY

The organization is even today highly decentralized and customer-focused as portrayed in Figure 1. The arrow shows how the organizational structure is designed to support customers:

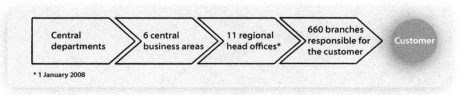

Figure 1: Handelsbanken Organization and Customer Focus

Handelsbanken describes its corporate philosophy in its annual review 2007:

- A strongly decentralised organisation- "the branch is the Bank"
- Focus on the customer-not on individual products
- Profitability is always given higher priority than volumes.
- A long-term perspective.

A RICH PICTURE OF THE COMPANY CULTURE

Having all this information and descriptions of how the bank thinks it is supposed to make money, the picture gets complicated even though the descriptions are quite simple and not hard to understand in reading. In order to try to understand what it all is about it could be helpful to draw a picture. Peter Checkland [Checkland, 1993] has developed a methodology to gather information about complex situations, where drawing a picture of the situation can be a good thing to do in the beginning of a process of understanding complex systems. A rich picture can be drawn without having to follow any particular rules and it could help one to see things more freely than if one starts to make a description of a situation in writing. It could help in detecting things that come into your mind but are quite intuitive. By drawing things that just come into your mind one can start seeing both objective and subjective aspects of things, connections, influences and so on.

So, I started by thinking of where the company culture really can be found and thought it to be within the employees, and it is there that my rich picture begins, namely with the employee standing there with all these influences from within and outside the bank. The rich picture is portrayed in Figure 2.

Figure 2: Rich Picture of Handlesbanken's Company Culture

Some Implications of the "Rich Picture"

Since descriptions about how the bank should be successful mostly are about people, their motivation and competence rather than financial or other "tangible" things, one starts thinking about how ten thousand individuals with their own viewpoints can be managed to work according to these principles? Since my (Susanna Göransson) field of research includes individual developmental and educational theories, individual variation is something that I cannot exclude from my mind when drawing the picture. Individual motives, experiences, life situation etc are not the same for any two employees of the bank. Since the principles of the company culture are described in a way that includes the individual, the overall company culture must be some kind of a sum of all these individuals working together in a very complex system. So I should really have to draw ten thousand of these pictures with all individual factors considered in order to understand the individual contributions of each employee to the overall company culture! Or can we look at the culture out of another perspective?

Svenska Handelsbanken has been highly competitive for many years, which can be related to a decentralized work organization within the framework of the company culture. Handelsbanken´s goal is to have higher profitability than the average of its competitors, which it has managed to do for the last 36 years. An explorative interview study of this bank [Wilhelmson et al., 2006] and a large survey in 2008 with all personnel in Sweden indicated very strong identification with this company culture amongst the personnel. There must therefore be some things in common for these ten thousand employees, some kind of value system that ties them together that could be considered a company culture and this is perceived somewhat in similar ways of most people in the bank!

THE COMPANY CULTURE AS A SYSTEM

Company culture has been especially in focus when the issue of how to build "learning organizations" has become a field of interest during the past two decades. Peter Senge has discussed the learning cycle and how a learning organization can become reality, and the set of beliefs and assumptions that develop in a learning organization are believed to be different from the ones in an ordinary hierarchical organization [Senge, 1990, Senge et al., 1994]. Senge stresses the guiding ideas of the organization and that these exist either they are deliberately developed or not. Ideas and truths that exist within the organization are guiding how people think, act and learn in their daily work. I will borrow Nonakas description [Nonaka, 1991].

> *"A company is not a machine but a living organism, and, much like an individual, it can have a collective sense of identity and fundamental purpose. This is the organizational equivalent of self knowledge—a shared understanding of what the company stands for, where its going, what kind of world it wants to live in, and, most importantly, how it intends to make that world a reality"*

So, these kind of deeper settings and beliefs can in Handelsbanken´s case either be deliberately designed and implemented or have just developed over time, or more likely, could be a mix of both of these. Clearly there were strong guiding ideas behind Wallander´s decentralization in the 70´s, but how are these ideas carried along until this day in the bank? This is a quite interesting part to look at, since the bank actually has made an effort to keep these ideas living and continually try to make all employees aware of these though mainly two tools, "Hjulet" (="The wheel") and "Mål och Medel" (=Goals and Means).

"Hjulet" (The wheel)

The "wheel" is the company yearly business planning process that includes action planning, individual follow-up, wage planning, and (career) development discussion as portrayed in Figure 3. Since it is a yearly process that comes again and again it can be seen as "wheel" and this is the description of the bank of the Wheel:

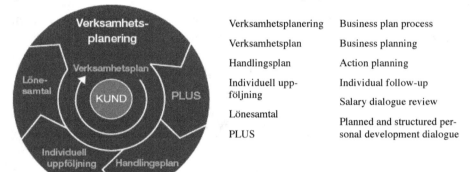

Verksamhetsplanering	Business plan process
Verksamhetsplan	Business planning
Handlingsplan	Action planning
Individuell upp-följning	Individual follow-up
Lönesamtal	Salary dialogue review
PLUS	Planned and structured personal development dialogue

Figure 3: The Wheel

"Mål och Medel" ("Goals and Means")

Mål och medel is a book that is renewed by each new CEO in the bank. It is a description of how the bank is supposed to work and achieve its goals. It describes values and beliefs that should guide employees of the bank and tries to clarify for the employee in what way my work does influence the success of the bank in total? In the recent version of Mål och medel the overall goal is described to be more profitable than the average of Swedish banks and the overall means to reach this goal are to lower costs and have more satisfied customers than the competitors.

A SYSTEMIGRAM OF "MÅL OCH MEDEL"

A Systemigram is a form of modeling created by John Boardman [Boardman and Sauser, 2008]. This modeling tool is especially designed to translate words and prose into graphical form, and this is something I think can be fruitful in trying to understand how "Mål och medel" is related to the existing values and beliefs of the bank. How are these beliefs supposed to lead to higher financial performance and can they be described as a part of a system that also includes financial elements? The construction process of a Systemigram is not described in this paper since it was described in Chapter 2 of this book, but one important guideline can be expressed as follows: *"To faithfully interpret the originally structured text as a diagram in such a way that with little or no tuition the original author, at the very least, would be able to perceive his or her writings, and additionally, meanings."* Figure 4 is a systemigram that captures the essential meaning of what the current CEO tries to relate in "Mål och medel", with the interpretation of how values, beliefs and financial success for the bank are connected as a system:

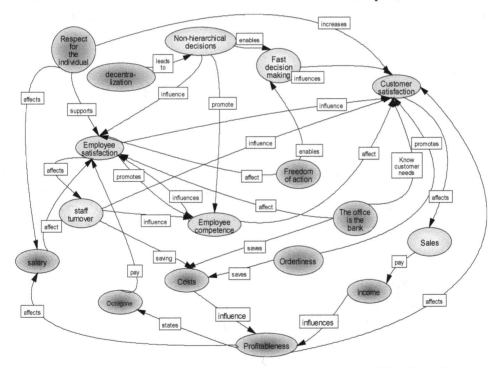

Figure 4: Systemigram of Elements and Relationships in Handelsbanken's Strategy (Note: Octogone is the name of the Employee Pension Fund)

Implications of the Systemigram

When working with the Systemigram some elements were more clearly described as important things in "Mål och medel", some are more vague but all are somehow present in the text. The elements that are easier to detect from the text are in the darker spheres and include values and beliefs of Handelsbanken, as well as financial issues. Elements that have showed themselves to me that were not actually present in my mind at the beginning and that are not clearly defined in "Mål och medel" as object or elements are mostly in the lighter spheres. The spheres "employee satisfaction", "employee competence" and "customer satisfaction" are the most interesting elements that gradually have pushed away elements that are more physical, for example the employee itself! Actually the employee is very present in "Mål och medel" but when the Systemigram developed, the physical employee turned to be of less importance than for example the competence that this employee is representing. The same things go for the Non-hierarchical decisions, Fast decision making, Sales and Staff-turnover spheres that really are things that are needed to understand how some elements are related. But these turned to be of such an importance that they got to be more than relations and are described as elements of their own importance in the Systemigram.

SYSTEMS OF INTEREST

Since the two methods used here turned out to mainly focus on the organization and not taking the outside world into account, it could be fruitful to just take a glimpse of how company culture is related to other systems and to its environment? Since company culture is a quite vague thing and one could have several different kinds of definitions of what it exactly is, I simplify somewhat and try to focus on what is around it rather than getting stuck in definitions at this point. This could be a description of Systems of Interest for the company culture of Handelsbanken:

- NarrowerSOI: the company culture, official and unofficial, guiding everyday work at Handelsbanken
- WiderSOI: how the work is done and organized, banking routines, everyday contacts between employees and with customers, "Mål och Medel"
- Narrow environment: demands of customers, rules and instructions within the bank, physical environment at the bank, IT systems, internal demands of profitability
- Wider environment: market situation, shareholders, Swedish Financial Supervisory Authority, competitors both in Sweden and abroad

CONCLUSIONS

Even though company culture, values, ideas etc are quite abstract things to identify and define, clearly a profit-making company as Handelsbanken considers these things important enough to be described and worked with. Today most of the understanding of the bank's company culture is described in a book named "Mål och medel" which tries to explain how values can lead to higher profitableness. My goal has been to in an even more explicit way understand the company culture as a system that ties together soft and hard issues, tangible and intangible things. Three methods have been used in this analysis and this has brought forward some new thinking about how this specific company culture works. Hopefully these tools can be helpful in trying to understand why some quite abstract values can, if processed in a strategic manner as Handelsbanken does, lead into a well-functioning and profitable and learning organization.

CASE STUDY REFERENCES

Boardman J and Sauser B (2008) Systems Thinking; Coping with 21th Century Problems, CRC Press, Boca Raton, FL.

Senge P (1990) The Fifth Discipline-The Art and Practice of The Learning Organization, Chatham, Kent: Doubleday, New York.

Senge et al (1994) The Fifth Discipline Fieldbook: Strategies and tools for building a Learning Organization, Currency Doubleday, New York.

Nonaka I (1991) The Knowledge-Creating Company, Harvard Business Review, Nov-Dec 1991.

Wallander J (2002) Med den mänskliga naturen-inte mot! Att organisera och leda företag, SNS Förlag, Kristianstad.

Wilhelmson L, Backström T, Döös M, Göransson S, Hagström T (2006) När jobbet är kul då går affärerna bra! : om individuellt välbefinnande och organisatorisk konkurrenskraft på banken, Arbetslivsrapport 2006:45, Arbetslivsinstitutet.

Chapter 3
Acting in Terms of Systems

As mentioned in the introductory chapter, thinking in terms of systems without acting to attempt to improve upon system situations is not very useful. Checkland by identifying "action to improve" in his Soft System Model is pointing to this need to act and then re-evaluate.While the nature of soft systems is different from that of hard systems, the idea of "engineering" structures, hard or soft, in order to achieve (or attempt to achieve) new behaviors is a common denominator as expressed in the discussion of the unification of disciplines in Chapter 1.The difference being that for hard systems, specific goals are established and then the system is engineered in order to achieve the verifiable goals.All of the system assets, be they physical, abstract or human activity based, that an organization and its enterprises define and deploy are "engineered" to achieve some purpose, mission or goal.These system assets when instantiated and deployed in responding to a situation must operate properly to provide their services.In this chapter we consider how these system assets are life cycle managed.

Problems and opportunities identified and described via the systems thinking analysis of models for both hard systems, soft systems and mixtures thereof are situation systems and thus are utilized for making decisions and can result in change actions.This will become evident as we now explore a paradigm for thinking and acting.

A PARADIGM FOR THINKING AND ACTING

A paradigm that will enable the reader to capture a broader view of systems and provide a useful means of continually focusing upon important aspects of thinking and acting in terms of systems is portrayed in Figure 3-1.The paradigm is, in fact, the result of combining two other known paradigms; namely the so-called OODA and PDCA loops.

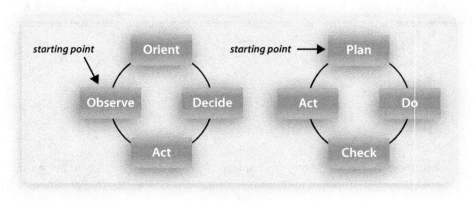

Figure 3-1: The OODA and the PDCA Loops

The OODA Loop

Col. John Boyd of the U.S. Air Force, a veteran of aircraft combat during the Korean conflict of the 1950s, set out to explain why some pilots succeeded in air combat while others, equally well trained, failed. Over the course of a lifetime, the insights that Boyd gained from studying dogfights grew into a broader description of tactical decision-making in dynamic situations [Boyd, 1987].Long taught in military circles especially for command and control, Boyd's ideas gained increasing influence in business and governmental circles.

The heart of Boyd's theory is based upon the premise that a tactical decision is the result of activities in a four-step loop.Boyd called these activities Observation, Orientation, Decision and Action.As noted in the figure, the loop begins with an observation.

Observation: Whether observation consists of the visual cues that guide a fighter pilot or the staff papers and briefings presented to a senior administrator, the decision-maker must first perceive and assimilate information about the environment as a basis for decision.

Orientation: Once the decision-maker has gained information through his/her or others observations, he/she must fit those pieces of information into a useful understanding of the situation.

Decision: The decision-maker selects a course of action.

Action: The desired course of action is executed.

The relationship of the OODA activities to the activities of science portrayed in Figure 1-1 is quite interesting.It is directly in line with the principles of systems thinking and in enterprises is typically executed continuously in leadership and management functions.This "spawning" for problems and opportunities is an integral part of the Change Model of Figure 1-13.

The PDCA Loop

The PDCA (Plan, Do, Check, and Act) loop was first introduced by Walter Shewhart in the 1920s as the activities required to achieve successful Statistical Quality Control.The PDCA concept was later popularized by MIT professor W. Edwards Deming as one of the guiding principles of TQM (Total Quality Management).Deming had worked under the mentorship of Shewhart at Bell Telephone Laboratories.

While there are some similarities as to the general goals of OODA and PDCA in respect to identifying problems and opportunities, the latter goes deeper into actually making changes, measuring the effect of changes and taking corrective actions to achieve planned goals.The description of the detailed meaning of individual PDCA activities has varied amongst numerous authors since its popularization by Deming. Some utilize PDSA (where the S means Study). When used in conjunction with soft systems, the use of study instead of check is quite appropriate.In the description that follows the meaning is related mainly to project management and has a direct coupling to the usage of PDCA within the scope of the Project related processes as provided in the ISO/IEC 15288 standard (described below).The loop begins with the creation of a Plan.

Plan: Create a project plan for accomplishing a goal or set of goals that are related to solving a problem or pursuing an opportunity.The plan will include the definition of the processes required to achieve the changes necessary to achieve the goals.

Do: Make the change.

Check: The results of the change are checked (verified) against the goals that were established.

Act: If necessary, corrective actions are taken to adjust the project plan, perhaps renegotiate the goals and then to recycle the loop until the goals are achieved or a decision is made to terminate the project.

There is a strong relationship between the PDCA activities presented here and the engineering treatment of structures and behaviors portrayed in Figure 1-1, especially when the project has the goal of designing and developing new structures, or altering existing structures.In general, the PDCA thinking is applied to all projects, for example, projects related to conceptualization, development, production, maintenance, or retirement of systems.

The application of OODA is always continuous in nature.The application of PDCA in guiding projects is discrete in nature; that is, it is typically applied for achieving specific goals within a specific time frame with provided resources after which the project is terminated.

OODA and PDCA Integration

The integration of the two loops is realized by coupling the Act activity of the OODA loop to the Plan activity of the PDCA loop.That is, the action to be performed involves the formation of a project, the first activity of which is the Plan. As a result of the execution of the project data and information concerning results, problems and opportunities are feed back for Observation in the OODA loop as portrayed in Figure 3-2.

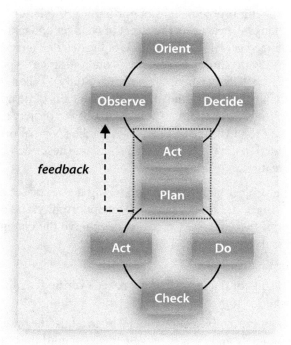

Figure 3-2: Integrating the OODA and PDCA Loops

The reader should keep this paradigm in mind as it relates to our mental model system-coupling diagram.The paradigm can be used to explain most any situation related to thinking and acting in terms of systems and relates directly to Situation Systems and Respondent Systems.It is a good model for understanding important relationships between enterprises and the projects that they create and monitor in order to manage change.Further, it is implicit in the project related processes provided by the ISO/IEC 15288 standard.The connection of the OODA part to the systems thinking methodologies described in the previous chapter should be obvious. We can observe that the systems thinking methodologies are providing support for the first three activities; namely Observing, Orienting and Deciding. While Checkland does indicate that action is to be taken, there is no specific guidance as to how to Act.In an enterprise setting, this most often will involve the initiation, execution and follow-up of projects.

REVISITING THE CHANGE MODEL

Now that the ground for thinking and acting in terms of systems has been established, it is useful to return to the Change Model introduced in Chapter 1 (Figure 1-13).An updated version appears in Figure 3-3.Consider the story that this figure conveys.

It is the Change Management function of an enterprise that makes decisions about required changes in systems.The OODA loop becomes a central driving paradigm for their operation.There is a definite need to focus upon the dynamic (emerging) behavioral aspects of multiple interacting systems (in the Orient activity).Thus, systems thinkers utilizing systems thinking languages and methodologies for gaining an understanding of existing dynamic situations can formulate a basis for decisions.From a strategic point of view it is important to be assured that the enterprise, via its system assets has the ability to create Respondent Systems in the form of Projects for handling the actual and potential situations that occur or (proactively) can occur.

In deciding upon the goal or goals of changes, the Change Management function (as the representative for the enterprise) operates as a line function.The execution of changes to operational parameters typically does not require a project organization and thus is directly ordered in accordance with vested authority and responsibility of operation and support line functions.In such cases, the Project Management aspect of Figure 3-3 may not be explicitly invoked.However, depending upon the scope of the change, it can be wise to follow the PDCA paradigm in order to accomplish effective operational change in a line organization.In either case, the response to situation coverage results in the formation of a Respondent System to meet the situations (real or thematic) as described in Chapter 1.

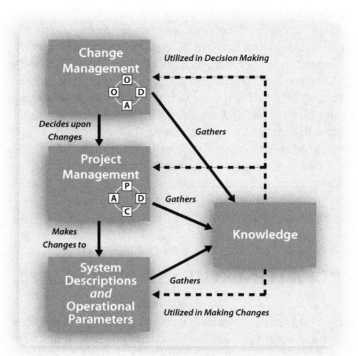

Figure 3-3: Change Model (Paradigm and Project Management)

For structural changes, the Change Management function defines, creates and monitors projects that carry out changes and verifies that the projects meet the goals.The PDCA paradigm, functioning as the control element for a Respondent System, is applied. Guided by the PDCA, Project Management deploys appropriate technical processes to make the required changes in system descriptions.

During the activities of Change Management, Project Management, and the execution of processes to accomplish the change, knowledge is gathered and becomes a part of the organization's (enterprise) human capital.As noted in Figure 3-3, the knowledge gained is fed back to assist in improving future decision-making and fed forward in improving capabilities in changing system descriptions and/or operational parameters.The reader should keep this revised Change Model in mind.

SYSTEMS ENGINEERING

"Systems Engineering is an engineering discipline whose responsibility is creating and executing an interdisciplinary process to ensure that the customer and stakeholder needs are satisfied in a high quality, trustworthy, cost efficient and schedule compliant manner throughout a system's entire life cycle."
Consensus definition provided by **INCOSE Fellows** [www.incose.org]

As with systems thinking, it is difficult to obtain a precise definition of the discipline of systems engineering.The definition provided above is based upon a consensus amongst peer members (Fellows) of the most prominent professional systems engineering organization.

The International Council on Systems Engineering (INCOSE) evolved during the early 1990s as a professional organization to promote the theory and practice of systems engineering.The discipline of systems engineering has been evolving since World War II.While early focus was placed upon complex hard systems, the focus has changed and successively dealt with soft systems as well.Furthermore, there are professionals who suggest that system thinking, as described in the previous chapter, is actually a part of systems engineering processes.

This book does not address the topic of Systems Engineering in depth. A number of books have been provided about systems engineering covering both architecting of systems as well as processes related to performing life cycle systems engineering activities [see, for example, Rechtin and Maier, 2003, Martin, 2000, Stevens, et.al., 1998, and Wasson, 2006].Further, INCOSE has produced a handbook for Systems Engineering where the principles and practices of the discipline are described [INCOSE, 2007].This body of knowledge is based upon the structure of the ISO/IEC 15288 standard and is utilized as a basis for the professional certification of systems engineers.The standard has been an attractive framework for expressing the processes and activities related to systems engineering.The following quote indicates why such a standard is important.

"Standards are generally required when excessive diversity creates inefficiencies or impedes effectiveness"
Hammond and Cimino [Hammond and Cimino, 2001]

The main point to keep in mind is that there is a converging body of knowledge and experience related to acting in terms of systems that is available as important guidance.

SYSTEM LIFE CYCLE MODELS AND PROCESSES

"Providing a basis for world trade in system products and services."

Requirements for the ISO/IEC 15288 standard, 1995.

Achieving this goal for the development of the world's first major domain independent systems standard was not an easy undertaking.The expert group that developed the standard worked for over six years in the preparation of ISO/IEC 15288:2002. The development of the standard has led to several important advances in understanding the nature of man-made systems, in formulating life cycle models and in the provision of a comprehensive set of system relevant processes. The main properties of the ISO/IEC 15288:2008 revised version of the standard are reviewed in this chapter. See Note 3-1.

Life Cycle Models

A vital aspect of being able to effectively develop, produce and operate as well as to make changes to systems is related to the definition and utilization of system life cycle models for the enterprise's portfolio of institutionalized systems assets as portrayed in Figure 1-6 and illustrated in Table 1-1.

The design and development of a System-of-Interest (i.e. the engineering part) transpires in one or only a few stages in the life cycle of the system.In fact, for most complex systems, design and development often represent a small part of the total effort as well as the total system costs.Thus, a vital part of thinking as well as acting in terms of systems involves understanding the structure of life cycle models and considering their deployment as guidance during the entire system life cycle.

The model presented in Figure 3-4 is a useful life cycle example.The System-of-Interest, provides its services and delivers its operational effect during the Utilization Stage.In parallel with this stage, a Support Stage is typically operated to assure that the system asset continues to provide its services and deliver its desired effect.

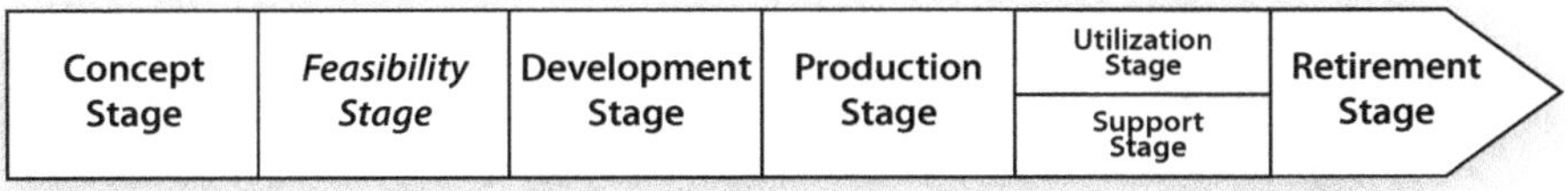

Concept Stage	*Feasibility Stage*	Development Stage	Production Stage	Utilization Stage	Retirement Stage
				Support Stage	

Figure 3-4: A System Life Cycle Model

The System-of-Interest began its life cycle journey in the Concept Stage where the needs of various stakeholders were identified and are most often transformed into some form of requirements specification.In this life cycle model, a Feasibility Stage is incorporated.Such a stage can be required when the System-of-Interest is very complex, or is an unprecedented (new type) of system where the structures that can deliver the required system services can potentially be designed and developed in a variety of manners.Based upon the requirements, one or more parallel endeavors (projects) are activated and their results evaluated in respect to critical trade-offs.As a result, a decision is made for the implementation approach and the relevant specifications are passed on to a Development Stage.

The Development Stage transforms specifications for the system product or service into a viable architecture design description that can be used in producing, operating, maintaining, and eventual retirement. The architecture design may be based upon the flow of materials, data, information as well as functions to be performed in transforming the flow into useful services.Alternatively, the architectural design may involve the use of interacting objects in which material, information and data are transformed and communicated to surrounding objects. The design of complex systems must also include the mechanisms necessary for the evolution of future versions of the system. That is, they must be designed for future structural as well as operational change.

The Production Stage utilizes the developed system description as a "template" to produce product (or service) instances.The system product becomes an asset to some party when it is put into operation during its Utilization Stage and supported by a parallel Support Stage.Logistic systems for repairs, spare parts, warehousing and delivery and help-desk systems for assisting operators/users are examples of typical forms of support.When the system is no longer required, or is to be replaced by a new version or another type of system, it is retired.

Models of life cycles, like the one illustrated in Figure 3-4, become instruments for managing the progression of change.The stages of the model define units of work that are to be accomplished and evaluated.Thus, the movement of a system through its life cycle is based upon life cycle related decisions made by the change management function of the organization or enterprise. The life cycle model provides decision points at which reviews of progress can be made and decisions related as to movement between stages as portrayed in Figure 3-5. Experience related to the definition of and utilization of life cycle models is an important part of enterprise Knowledge intellectual capital.

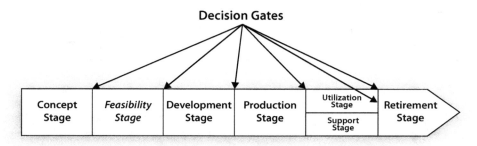

Figure 3-5: Life Cycle Model Instrumented with Stage Decision Gates

Enabling Systems

A major contribution of the ISO/IEC 15288 standard in respect to achieving a broad view of the life cycle management of systems is the definition of and systematic treatment of enabling systems.Enabling systems contribute to the achievement of the goals of various stages as the System-of-Interest moves through its life cycle. Thus, returning to the life cycle model example, Figure 3-6 illustrates the utilization (deployment) of enabling systems at various stages in the life cycle.

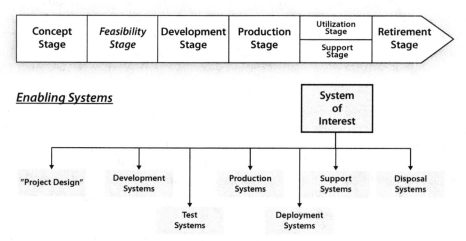

Figure 3-6: The Deployment of Enabling Systems

In this example, when the System-of-Interest, as a product instance, is in its Utilization and Support Stages, Support systems providing, for example, Logistics and Help-Desk functions assist in the operation and maintenance of the system.When in the Production Stage, a variety of Production Systems are deployed.During the Feasibility and Development Stages, Development and Test Systems are utilized.A useful enabling system to deploy during the Concept Stage is a Project Design System that assists in identifying the projects required for moving the System-of-Interest through its life cycle in terms of System Breakdown (SBS) and/or Work Breakdown (WBS) Structures. Deployment Systems can be used to facilitate the transition of system product or service instances into operative assets.Finally, in the Retirement Stage, Disposal Systems are deployed in order to enable proper termination of the instances of or the entire System-of-Interest.

The ISO/IEC 15288 standard is reapplied for each System-of-Interest at each recursive level of a system.Further, the standard can be applied to each enabling system when it is treated as the System-of-Interest.This approach assures that parties responsible for change management and parties performing transformations on system descriptions must think about and plan for the availability and utilization of enabling systems at all points in the life cycle.Thus ISO/IEC 15288 makes a significant contribution to seeing wholes and promotes thinking and acting in terms of systems.

Note: The complex of a Narrow System-of-Interest (NSOI) and all of its Enabling Systems as part of the Wider System-of-Interest (WSOI) forms a larger system with potentially complex relationships. When problem or opportunity situations arise in the management of systems, it can be necessary to examine the holistic properties according to the methodologies of systems thinking that have been described.Many of the problems can be due to non-technical factors that are best studied as thematic systems that span over a range of institutionalized system assets as portrayed in Table 1-3.

Defining Stages

The ISO/IEC 15288 standard specifies the structure of an individual stage, but does not dictate any specific multi-stage model.Multi-stage life cycle models like the one presented in Figures 3-4, 3-5 and 3-6 are based upon the needs of the enterprise for managing the various types of systems in their portfolio.

The requirement for a stage building block is that it includes the purpose of the stage, as well as the stage outcomes. The generic life-cycle model described in ISO/IEC 24748-1 [ISO/IEC 24748-1, 2009] composed of Concept, Development, Production, Utilization, Support, and Retirement stages is put forth as an illustration.Each stage is described by an overview followed by the stage purpose and the typical stage outcomes.

To illustrate the description of a stage, consider the following possible definition of a Feasibility Stage in a manner that corresponds to the requirement for stage descriptions in the standard.

Feasibility Stage

Overview

This stage can be viewed as an extension to the Concept Stage. A Feasibility Stage is utilized in situations where significant uncertainty exists as to whether a viable, risk and cost effective solution can be established for further development into a desired System-of-Interest that meets required needs and can provide required services.

In striving to establish a basis for development, multiple approaches to viable solutions are to be explored via projects. In such cases, it is important that each project result be reported upon and evaluated in respect to risks, cost effectiveness, estimated time schedule, and other factors that are deemed to be of importance for the type of System-of-Interest being considered.

Feasibility Stage Purpose

The Feasibility Stage is executed to determine whether a viable, risk and cost effective approach to implementing the required System-of-Interest can be established based upon one or more potential solutions.

Feasibility Stage Outcomes

The outcomes of the Feasibility Stage are listed below:

- One or more potential solutions are identified and described.

- For each potential solution, risk identification, assessment and mitigation plans for this stage and succeeding stages are described.

- For each potential solution, an identification and specification of services needed from enabling systems throughout the life cycle.

- For each solution, a concept for execution of succeeding stages is described.

- For each solution, plans and exit criteria for the Development Stage are described.

An effective approach to getting started with developing a life cycle model is to utilize the illustrative descriptions from the life cycle guidance provided in [ISO/IEC 24748-1, 2009].Then by modifying the individual stages descriptions, adding and/or removing stages, one may develop appropriate life cycle models to meet the needs of each type of System-of-Interest that is to be life cycle managed by an enterprise.

In practice, it is expected that an enterprise will develop life cycle models for a small number of types of systems that are incorporated in its system portfolio. In an extended enterprise environment in which many organizations are involved agreement between partners on stage models will promote the cooperation that is necessary for achieving common purposes, goals and missions.

The ordering of the stages presented in Figures 3-4, 3-5 and 3-6 should not be taken to mean that stages are executed sequentially.In fact, for all but simple systems, sequential execution would be an exception.It is very common that iterations of stages are performed particularly in refining concepts and developing solutions, possibly through the development of prototypes.Such iterations can result in a formation of an incremental development model.Further, the progression through stages may be partitioned according to the desire to acquire a product and/or service in increments over time (so-called incremental or progressive acquisition).Thus, it is the enterprise, via initiating stages and the evaluation of results in subsequent decision-making that determines the stage execution order of the life cycle model. In Chapter 6, the life cycle management of various types of systems including the use of various development models is explored in further detail.

Processes for System Actors

There are a variety of actors involved in various aspects of systems during their life cycle.To meet the needs of the various actors, the twenty-four process descriptions provided in the ISO/IEC 15288:2008 standard are divided into four categories as portrayed in Figure 3-7.

In the standard, each process is described in terms of a process *purpose*, process *outcomes*, process *activities* and process *tasks* to be used in achieving the outcomes.When an enterprise decides to apply a process, it can use the process as described and thus will be compliant with the requirements of the standard.However, most often, a process description provides a starting template from which modifications, extensions and deletions are made in order to construct a process that meets the enterprise needs.Such alterations are called *tailoring* as described later in this chapter as well as in Annex A of the standard.

In the following, descriptions of the categories of processes aimed at satisfying the needs of particular system actors are provided as well as a summary of the purposes of the processes within each category.

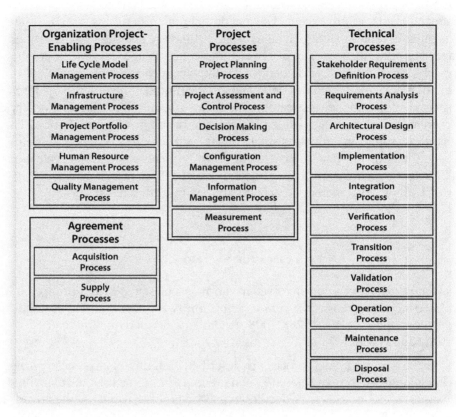

Figure 3-7: The ISO/IEC 15288:2008 Processes

Organization Project-Enabling Processes

These processes are provided to manage an enterprises capability to acquire and supply system products or services through the initiation, support and control of projects. They provide the resources and infrastructure necessary to support projects and ensure the satisfaction of organizational objectives and established agreements.

In practice the processes of this category are most often assigned to some form of enterprise entity that has continuous line management responsibility for the systems of the enterprise system portfolio.The application of these processes provides an environment in which the OODA (Observe, Orient, Decide, Act) loop of Change Management described earlier in this chapter can be applied in a rational manner. The purpose of each process is as follows:

Life Cycle Model Management	define, maintain, and assure availability of policies, life cycle processes, life cycle models, and procedures for use by the organization with respect to the scope of the International Standard
Infrastructure Management	provide the enabling infrastructure and services to projects to support organization and project objectives throughout the life cycle
Project Portfolio Management	initiate and sustain necessary, sufficient and suitable projects in order to meet the strategic objectives of the organization
Human Resource Management	ensure the organization is provided with necessary human resources and to maintain their competencies, consistent with business needs
Quality Management	to assure that products, services and implementations of life cycle processes meet enterprise quality goals and achieve customer satisfaction

Agreement Processes

The processes define the activities necessary to establish an agreement between two enterprises. The Acquisition Process provides the means for conducting business with a supplier of system products that are supplied for use as an operational system, of services in support of an operational system, or of elements of a system being developed by a project. The Supply Process provides the means for conducting a project in which the result is a system product or service that is delivered to the acquirer.

In practice these processes, in addition to being utilized in inter-enterprise agreements can be used at any level within an enterprise; for example, in formulating agreements between projects or other entities within the same enterprise. The purposes of these two extremely important processes are as follows:

Acquisition	obtain a product or service in accordance with the acquirer's requirements
Supply	provide an acquirer with a product or service that meets agreed requirements

Project Processes

These processes are utilized to establish and evolve project plans, to assess actual achievement and progress against the plans and to control execution of the project through to project retirement. The processes collectively or individually may be invoked at any time in the life cycle and at any level in a hierarchy of projects as required by project plans or unforeseen events. The processes are to be applied with a level of rigor and formality that is dependent upon the risk and complexity of the project.

The first two processes are dedicated to planning, assessment and control and are key to sound management practices.They are directly related to the PDCA (Plan, Do, Check, Act) loop described earlier in this chapter.Planning, assessment and control are evident in the management of any undertaking, ranging from a complete organization down to a single life cycle process and its activities. The purpose of each process is as follows:

Project Planning	produce and communicate effective and workable project plans
Project Assessment and Control	determine the status of the project and direct project plan execution and ensure that the project performs according to plans and schedules, within projected budgets and it satisfies technical objectives
Decision Making	select the most beneficial course of project action where alternatives exist
Risk Management	identify, analyze, treat and monitor the risks continuously
Configuration Management	establish and maintain the integrity of all identified outputs of a project or process and make them available to concerned parties
Information Management	provide relevant, timely, complete, valid and, if required, confidential information to designated parties during and, as appropriate, after the system life cycle
Measurement	collect, analyze, and report data relating to the products developed and processes implemented within the organization, to support effective management of the processes, and to objectively demonstrate the quality of the products

In addition to the two central project management processes of planning, assessment and control, there are five processes that are used in supporting the work of projects. However, other system actors may also benefit from utilizing these processes.The Decision Making process in addition to its usage in projects forms the basis for tailoring and utilization in evaluating Decision Gates as described in this chapter and as illustrated in Figure 3-5.In some enterprises the Risk Management, Configuration Management, Information Management and Measurement processes may be centralized and thus applied at an enterprise level.

While the standard provides all of the essential rudiments of Project Management there are additional sources of knowledge that provide further guidance.For example, the PMI (Project Management Institute) publishes extensive guidance in this area. [PMI, 2008]

Technical Processes

These processes are utilized to:

- define the requirements for a system,
- transform the requirements into an effective systems product,
- permit consistent reproduction of the product where necessary,
- use the product to provide the required services,
- sustain the provision of those services and
- dispose of the product when it is retired from service.

The processes define the activities that enable enterprise and project functions to optimize the benefits and reduce the risks that arise from technical decisions and actions. These activities enable products and services to possess the timeliness and availability, the cost effectiveness, and the functionality, reliability, install-ability, maintainability, usability, resilience, and other qualities required by acquiring and supplying organizations. They also enable products and services to conform to the expectations or legislated requirements of society, including health, safety, security and environmental factors.

In practice projects select all or a subset of these processes in order to perform the work required in progressing systems through their life cycle.Thus application of the processes within a project corresponds to the Do part of the PDCA loop described earlier in this chapter. The purpose of each process is as follows:

Stakeholder Requirements Definition	define the requirements for a system that can provide the services needed by users and other stakeholders in a defined environment
Requirements Analysis	transform the stakeholder, requirement-driven view of desired services into a technical view of a required product that could deliver those services
Architectural Design	synthesize a solution that satisfies system requirements
Implementation	realize a specified system element

Integration	assemble a system that is consistent with the architectural design
Verification	confirm that the specified design requirements are fulfilled by the system
Transition	establish a capability to provide services specified by stakeholder requirements in the operational environment
Validation	provide objective evidence that the services provided by a system when in use comply with stakeholders' requirements achieving its intended use in its intended operational environment
Operation	use the system in order to deliver its services
Maintenance	sustain the capability of the system to provide a service
Disposal	end the existence of a system entity

When the Implementation Process is invoked to produce a system element, it implies the application of domain specific knowledge and standards that are needed in order to supply the type of system element that is required.Note that this can be a recursive reapplication of ISO/IEC 15288 when the element to be provided is to be treated as another System of Interest.

The operation and/or maintenance of a system as an enterprise asset may be performed in project form or via continuous system operation authority and responsibility assigned to a line organization.

DISTRIBUTION OF LIFE CYCLE EFFORT

It is important to observe that the execution of process activities are not compartmentalized to particular life cycle stages.Successful life cycle management for complex systems is dependent upon the interaction of the various actors executing process activities that contribute to a holistic view of the system.A useful manner of portraying this is through a hump diagram [Kruchten, 2003] as provided in Figure 3-8.This particular hump diagram was provided by Rick Adcock and is directly related to the ISO/IEC 15288 processes and generic life cycle model presented in this chapter.

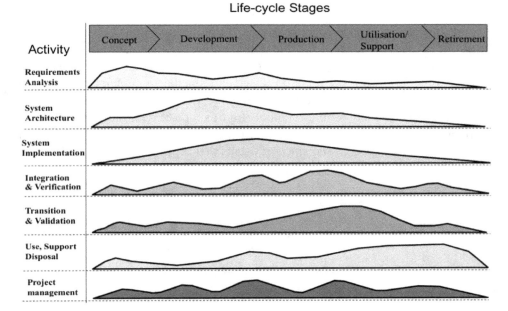

Figure 3-8: Hump Diagram of Life Cycle Activity (Rick Adcock, personal communication)

The lines on this diagram represent the activity intensity in the process areas over the life cycle. The peak in activity represents the point in the life cycle when a process becomes the main focus of the current stage. The life cycle stages along the top of the diagram are purely illustrative and not drawn to scale.The activity before the peak may represent either (a). through life issues raised by a process, e.g. how will system maintenance or retirement constraints be dealt with in the system requirements or system design or (b). forward planning to identify the resources need to complete future process activities, e.g. identifying and planning for system trials or tests.

For most systems, the Utilization/Support stages would likely be by far the longest part of the lifecycle. Requirements Analysis has a large input during the concept stage, but requirements are refined, reviewed, and reassessed over the rest of the life cycle. Similarly, Integration and Verification is conducted during the transition from Development to Production. This is only possible if Integration and Verification issues, strategies, and risks are considered in earlier stages.As noted, Project Management processes are executed during the entire life cycle. This may be the result of a single project or multiple related projects executed during the life cycle.

AN EXAMPLE OF STANDARD UTILIZATION

In order to gain a broader perspective of the utilization of ISO/IEC 15288, it useful to examine a model of the dynamic relationships that can evolve from the interaction of executing the processes from the four process categories (see Figure 3-9).This model portrays a situation in which an acquiring enterprise forms an agreement with a supplying enterprise for delivery of products or services.While the model conveys many of the essentials for this type of relationship, it must not be interpreted as representing all possible modes of standard utilization, nor is to be interpreted as a specific sequence of events that occur during the relationship.

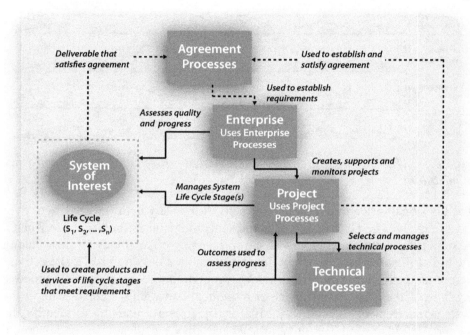

Figure 3-9: A Model for the Acquisition of System Products and Services

As described earlier, the standard is applied to one system at a time; namely, the System-of-Interest which is denoted in the model as having a life cycle composed of stages $(S_1, S_2, \ldots, S_n)$.In this example, an acquiring enterprise that has a need for a System-of-Interest product or service (or for the products or services delivered as a result of executing one or more life cycle stages) seeks to establish an agreement with a supplying enterprise.

Given the need of the acquirer as the starting point, observe the following potential dynamic aspects of the acquirer-supplier relationship:

a. The acquirer as a part of the acquisition process establishes requirements to be agreed upon with a supplying enterprise.

b. The supplier creates and monitors a planning project that is charged with responsibility to provide the product or service according to the requirements.

c. The project selects technical processes that are appropriate to fulfilling the requirements.

d. The enterprise and project uses planning information in establishing an agreement with the acquirer (perhaps a formal contract).

e. Projects are established to manage the work to be done in one or more life cycle stages.

f. The technical processes of the projects are executed in order to create products and services of the life cycle stages.

g. The suppliers project management utilizes the outcomes from execution of the technical processes in order to assess progress.

h. The results of life cycle stage related projects are used by the enterprise as stage decision gates in order to assess quality and progress.

i. The products and services (and any other deliverables) are supplied to the acquiring organization in completing the agreement.

As with all models, there are many aspects that are not portrayed, however, the described relationships provide a useful context for how the standard can be applied in an actual situation.

TAILORING TO SPECIFIC NEEDS

An important aspect of the standard is the ability to tailor the processes to meet specific needs as indicated in the following:

Tailoring	adapt the processes of the international standard to satisfy particular circumstances that:
	• surround an organization employing the standard in a agreement;
	• influence a project that is required to meet an agreement in which the standard is referenced;
	• reflect the needs of an organization in order to supply products or services.

The tailoring process can result in a viable multi-stage life cycle model. Alternatively individual life cycle stages that influence the fulfillment of an agreement to supply a product or service can be defined. In doing the tailoring new or modified processes can be defined.

Some of the factors that can be taken into account in tailoring are the stability and variety of operational environments, commercial or performance risks, system novelty, size or complexity, time schedules, issues of safety, security, privacy, usability, and availability as well as opportunities arising from emerging technologies.

In practice, both the process descriptions provided in the standard as well as the illustrated life cycle model in [ISO/IEC 24748-1, 2009] can be used as guidance in establishing both tailored processes and appropriate life cycle models.

ANOTHER LOOK AT THE CHANGE MODEL

In Figure 3-3, the basic Change Model was extended to reflect the OODA and PDCA paradigms as well as to reflect the relationship of Change Management and Project Management. Now that the main features of the ISO/IEC 15288 standard have been considered, it is important to relate the standard to the Change Model as portrayed in Figure 3-10.

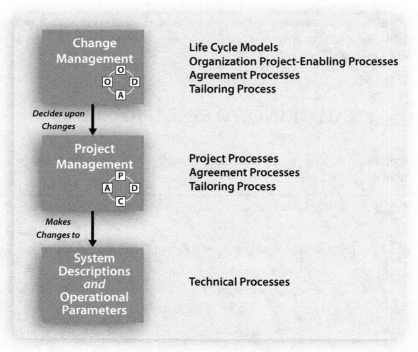

Figure 3-10. Change Model (Usage of ISO/IEC 15288)

The Change Management function as a representative for the enterprise deploys Life Cycle Models for establishing and controlling projects that make structural changes to system descriptions.The Organization Project-Enabling Processes provide the necessary environment in respect to policy, investment decisions, life cycle model and process maintenance, resource provision and quality management. Within this environment Change Management functions operate on a continuous basis according to the OODA loop.In acting to make changes, a Change Management function can utilize the Agreement Processes to establish a relationship (at an appropriate level of formality) with other enterprises or within the same enterprise.Finally, the Change Management function can utilize the Tailoring Process in providing appropriate processes and life cycle models.

Project Management is supported by core Project Processes for planning, assessment and control that correspond to the P, C, and A activities of the PDCA paradigm.Further support is provided for decision-making, risk management, configuration management, information management and measurement.It is also possible for projects to take advantage of the Agreement Processes in formulating agreements with other projects in the same enterprise.Formalization of these relationships is vital in avoiding misunderstandings amongst projects charged with cooperating in achieving a collective result.A project may also utilize the Tailoring Process to tailor processes in the international standard or further tailor processes established by an enterprise in order to deal with project specific needs.

The Technical Processes are employed for providing a first version of or making changes in System Descriptions and are executed within the scope of a project as the D part of the PDCA paradigm.Operation and Maintenance Processes can be incorporated within the scope of a project, but are often executed by a line organization that has authority and responsibility for the utilization related processes.Such an organization is also involved in making changes to Operational Parameters.

In the following chapters where the particulars of the Change Management Model are explored in further detail, the process purpose descriptions from this chapter are re-inserted at appropriate points in order to provide clarity of the context of their utilization in the implementation of change management.

KNOWLEDGE VERIFICATION

1. How can the OODA and PDCA loops be related to the line management and project management activities in your organization?

2. Explain how the revised Change Model can be applied in your organization, its enterprises, and its projects.

3. Identify some typical operational parameter decisions and structural change decisions made in your organization.

4. Once again, make sure that the terminology System-of-Interest, System Element, Enabling System, and Recursive Decomposition are understood.

5. Relate the life cycle of a system with which you are familiar to the system life cycle exemplified in this chapter.

6. Building upon the life cycle identified in exercise (5), define what enabling systems were necessary to support the system in its life cycle.

7. What is a decision gate and how is it used by an enterprise?

8. How do the life cycle stages Concept, Development, Production, Utilization, Support, and Disposal match the type of systems that are vital for your organization?

9. Match the process categories, Organization Project-Enabling, Agreement, Project and Technical to the structure of your own organizations.Given the purposes provided, identify whether these processes (or similar processes) are performed and where they are performed in your organization.

10. Are you convinced that the Technical Processes represent a complete set for creating, operating, maintaining and disposing of all types of man-made systems?

11. Compare the distribution of effort portrayed in the hump diagram with experiences from system related projects in your organization.

12. What is tailoring? How do you think tailoring of processes in the standard could be accomplished in your organization?

Note 3-1: This book is not intended to, nor is permitted to present the detailed guidance provided by the ISO/IEC 15288 standard.It does however introduce and treat the broad scope of the standard at a level of detail sufficient to understand the application of the concepts and principles in all essential aspects of the management of system life cycles.Thus, the material presented in this book complemented with the standard provides an excellent means of getting started in applying ISO/IEC 15288 in various systems related situations.

Further, the original publication of the standard was in 2002. In 2008, an updated version has been published in an effort towards unifying systems and software engineering standards.However, the guiding concepts and principles of the standard have not been altered.The presentation in this book is based upon the revised version that is ISO/IEC 15288:2008.

Chapter 4
System Descriptions and Instances

A system is a system in the eyes of the beholder
(The kaleidoscope view)

When is a system a system? As noted in Chapter 1, it is essential to have a proper perspective in this vital question. The Systems Survival Kit has provided guidance to this question. An institutionalized asset or respondent man-made System-of-Interest arises when two or more system elements have the "togetherness" property and thus are related in a defined manner in order to achieve a purpose, meet a goal, and/or accomplish a mission. Further, such Systems-of-Interest when instantiated and via emergent behavior meet a need, offers services and delivers effect that could not be provided by any single system element. Systems based upon the system elements to be integrated and the relationships between the elements can be described at various stages of their life cycle, in a number of manners. This includes textural, tablular and graphical descriptions as well as a variety of domain and discipline related notations and languages. Further, as noted in Chapter 1, instances, produced based upon system descriptions, in the form of system products and/or services are operated/utilized to provide desired effects. In this chapter, essential properties of system descriptions and instances that arise during life cycle management of systems are examined in further detail by building upon the knowledge provided in Chapters 1 to 3.

LIFE CYCLE TRANSFORMATIONS

In the previous chapter, we considered a set of processes that are used to perform systems related work (a part of acting in terms of systems). Various work products are produced as a result of "executing" carrying out processes during the life cycle as the System-of-Interest evolves from need to concept and to reality in the form of products and services. To portray these transformations based upon the knowledge that has been provided thus far, consider the life-cycle structure illustrated in Figure 4-1.

Here we observe at the top of the figure that the System-of-Interest is first described as Defined Abstract Systems that are then transformed into concrete Defined Physical and/or Human Activity Systems when they become a product, that is instantiated for utilization. An eventual retirement of a System-of-Interest involves disposing of instances and can also involve retirement of the system definition, that is, the Defined Abstract System.

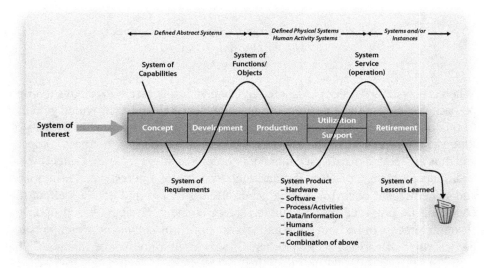

Figure 4-1: Life Cycle Transformations (System-of-Interest Versions)

It is important to note the perspective portrayed in the figure in naming the various stage and process related work products as "systems". We view the various descriptions as well as the eventual product as "versions" of the System-of-Interest. That is from a need, the first version of the System-of-Interest is created as a System of Capabilities. This description meets all of the criteria for a system as defined in Chapter 1 where the most fundamental criteria is the property of "togetherness".

From this System of Capabilities, the next version of the System-of-Interest is created in the form of a System of Requirements reflecting both the functional as well as non-functional requirements to be placed upon the System-of-Interest. The next version of the System-of-Interest is a System of Functions or Objects that describe the basic transformations that the instantiated System Products are expected to perform when they provide their service. Typically, this involves some type of flow of energy, material, data or information.

In order to provide for orderly development, production and usage, it is important to keep consistency between the various descriptions, that is, traceability between the elements of the various versions of the System-of-Interest.

Based upon the description versions, System Products are produced as the result of the integration of elements that can include hardware, software, processes/activities, data/information, humans, facilities, natural elements or combinations thereof. When the product is utilized in its final environment, it provides the System Service, that is the behaviors that it has been designed to achieve.

One further version of the System-of-Interest that is most often forgotten is to capture information about the history of the System-of-Interest in the form of a System of Lessons Learned based upon system conception and development as well as product instances and the services they have provided.

The reader should keep this "system perspective" on life cycle transformations in mind as we now proceed to consider various aspects of life cycle transformations.

KEY SYSTEM SUCCESS FACTORS

The development and utilization of successful systems requires a holistic life cycle approach. Your author has been involved in many system efforts for over 50 years and as a result, has collected a number of key success factors that have become personal lessons learned in pursuing holistic approaches.

Driving Concepts and Principles

Humans are driven by what they understand. When things get too complex, they either give up or attempt to find some guiding concepts and principles to which they can attach meaning and utilize in coping with complexities. Thus, the identification of a small number of central "driving" concepts and principles is an essential aspect of the early stages of the life cycle where the abstract description systems are established. Remember in Chapter 1 concrete semantics for man-made systems based upon concept definitions and related principles were presented as a part of the Systems Survival Kit. Having observed the power of this small number of driving concepts and principles for understanding the essential properties of systems, let us now examine why concepts and principles are so important for system development and utilization. To gain a deeper understanding or what is meant by concepts and principles, consider the following definitions provided by [Lawson and Martin, 2008].

A concept is an abstraction; a general idea inferred or derived from specific instances. For example, we look at our pet dog and we can infer that there are other dogs of that type. Hence, from this observation (or perhaps a set of observations) we develop a concept of a dog in our mind. Concepts are bearers of meaning, as opposed to agents of meaning and can only be thought about, or designated, by means of a name. For example:

- Gothic, Romanesque, Victorian, Baroque
- Box girder, Cable stayed, Cantilever, Clapper, Pontoon, Draw
- Object Oriented, SOA, Message Based
- Time Driven, Event Driven, Synchronous, Asynchronous

A principle is basically a rule of conduct or behavior. To take this further we can say that a principle is a basic generalization that is accepted as true and that can be used as a basis for reasoning or conduct. [WordWeb.com] A principle can also be thought of as a basic truth or law or assumption. [ibid.] Principles depend on concepts in order to state this truth. Hence, principles and concepts go hand in hand; principles cannot exist without concepts and concepts are not very useful without principles to help us understand the proper way to act.

As portrayed in Figure 4-2 explicitly stated and communicated concepts and principles are guiding factors in providing the know-how to achieve the purpose, goals and missions of a system. But, most importantly, when treated properly they become the catalyst for achieving stability via consistent decision making during the life cycle of the system.

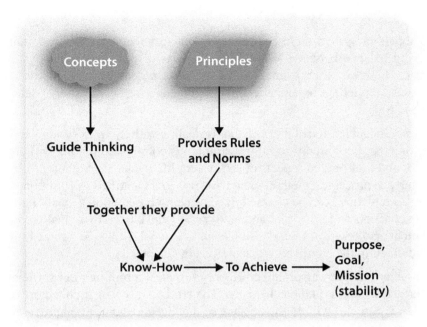

Figure 4-2: The Central Role of Concepts and Principles

To illustrate the vital role of driving concepts and principles, consider the following example:

Some Concepts	Some Principles
Farms	
Cattle	Cattle contained by fences
Fence	Fences must be maintained
Containment	Containment contributes to a stable farm environment

Farmers that understand and abide by the Concepts and Principles make consistent decisions and can maintain a stable farm environment. Farmers that do not abide by the concepts and principles spend most of their time chasing cows.

Unfortunately, there are many examples of cow chasing that can be observed in the world of complex systems. The development of systems-related concepts and principles is based upon a deep understanding of a particular system's environmental context, needs and requirements, the critical properties and characteristics of the system and the potential elements of the solution space that lead to a desirable architecture description.

Role of Architecture

In establishing system architectures it is vital to understand the importance of achieving balance between multiple key aspects of developing the System-of-Interest. These key aspects include architecture, processes, methods and tools, models and modeling, organization, and competence as described by [Bendz and Lawson, 2001].

A significant portion of the problems in dealing with complex systems is related to the fact that poor architectural underpinnings complicate virtually all process, method, and tool-related aspects of the system life cycle. This relationship of architectures to processes, methods, and tools was first identified by [Lawson, 1994]. Strong architectures can be characterized as being based upon a small number of driving concepts and principles and are typically developed by a small number of competent people with a shared vision leading to what is called organized simplicity as defined in the complexity discussions in Chapter 1.

In addition, a strong architecture only provides a minimal but sufficient set of interface standards and mechanisms. The effect of a strong architecture as portrayed in Figure 4-3 will tip the balance so that dependencies upon processes and related methods and tools become lighter. On the other hand, a weak architecture will tip the balance in the opposite direction so that processes, methods, and tools become heavy in order to compensate for weak architecture complexities. Unfortunately, since the mid-1970s, there has been a continued growth in complexities of computer-based systems leading to an emphasis upon heavy processes, methods, and tools. That is, emphasis has been placed upon how to do the job instead of doing a good job. A conclusion that can be drawn from this relationship is that it is worthwhile to expend efforts to develop proper architecture underpinnings.

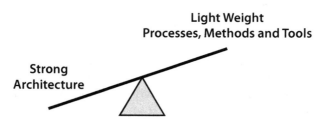

Figure 4-3: Balancing Architecture with Processes, Methods and Tools

Role of Processes

Even with strong architectures, the complexity of modern systems certainly requires the definition and utilization of well-defined processes that are appropriately allocated to the actors involved in achieving system related organization/enterprise objectives. Thus, the balance portrayed in Figure 4-4 is also important. A strong approach to planning, executing, assessing, and controlling the processes utilized in system related projects lightens the burden of the organization in achieving system related goals. Further, it reduces the reliance upon the competencies of one or a few system heroes. On the other hand, a weak approach to processes as illustrated typically results in attempts to reorganize organizations in their search to satisfy enterprise goals. Further, the lack of a structured process approach within a complex organization is a breeding ground for the heroes approach to achieving system goals. Most typically, reliance upon heroes in some stages of the life cycle (typically development) creates significant risks and eventual cost explosion in respect to the latter stages of the system life cycle. Many maintenance nightmares have been created due to the reliance upon heroes.

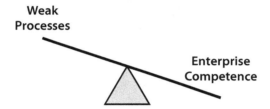

Figure 4-4: Balancing Processes with Enterprise Structure and Competence

In summary, a holistic approach that takes into account all key aspects and achieves an appropriate balance is a key ingredient to architecting, implementing using and maintaining successful systems. Given these general success factors, let us now examine some important aspects of the life cycle.

IMPORTANT ASPECTS OF THE LIFE CYCLE

The various system version transformations that take place during the life cycle as portrayed in Figure 4-1 form a basis for defining the authority and responsibility of system actors. In order to focus the discussion of important aspects of the transformations, we introduce three fundamental transformations; namely Definition, Production and Utilization as portrayed in Figure 4-5. Even though a given life cycle stage structure may vary from the general format portrayed in Figure 4-1 these three transformations are generic for all types of man-made systems.

Further, in the figure the utilization of the universal System-Coupling Diagram model is provided in order to portray its applicability in the "situations" that arise during the life cycle management of systems.

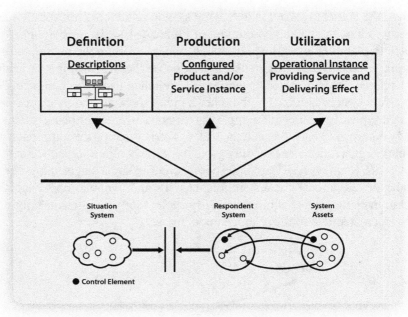

Figure 4-5: Fundamental Life Cycle Transformations and the System-Coupling Diagram

A Collection of Situation-Respondent Systems

As work proceeds during the life cycle, the "situation" changes. Thus, we begin with the situation that there exists a need for a system and the desire to create a system that will meet those needs. In the early stages of the life cycle the expression of need is transformed into a System of desired Capabilities. This transformation is the result of some form of organized effort, typically a project. In performing the work of the project relevant System Assets are taken from the Enterprise portfolio and collected in a Respondent System (in this case a project) that has as its objective (goal) to produce an end-state of a Situation System where the needs have been transformed into a System of Capabilities.

The System of Capabilities is the first description of the System-of-Interest. In another project or within the scope of the same project, the System of Capabilities is used as input to a Respondent System (project) now armed with System Assets necessary to do the transformation from the System of Capabilities to the

System of Requirements. This end-state results in another description version of the System-of-Interest. Then via another project or the same project, the System of Requirements is used as input to the Respondent System (project) that performs the work to transform the earlier two descriptions into a further description as a System of Functions/Objects where most typically the flow of energy, materials, data/information and transformations that will be accomplished by system instances are described. Thus, collectively the descriptions produced have provided a Situation System that contains as elements, the successive definitions of the System-of-Interest.

The definitions provided are utilized as the "blueprints" in producing one or multiple instances of the system. Thus production begins with a situation where the System-of-Interest including all of its elements and their relationships has been defined. Like the previous application of the System Coupling Diagram production (as a line activity or project) forms a Respondent System based upon the System Assets needed for production. For some systems, Production involves physical elements according to integration specifications, for others it assembles integrating elements in the form of people, software, processes, data and information. For systems containing both technical and non-technical system elements, it will involve both forms of integration. The end-state of the Situation System is the production of one or multiple instances of the System Product. The Production stage can continue as long as needed in order to meet customer demand for the System Product.

The provisioning of operational instances of the System Product to customers (users) establishes a new System Asset for the customer's portfolio. These assets are then utilized when needed as elements in Respondent Systems that are relevant for the customers needs.

The utilization of some types of defined physical System Assets will result in depletion or destruction of the instance whereas for other types of system, the instance remains and is available for continuation of the provision of services. For some systems the sustained utilization of instances is dependent upon various forms of support in the form of logistics for spare parts and maintenance as well as in some cases help desks. The final retirement decommissioning (disposal) of a system can result in reclaiming physical elements of instances and/or the archiving of information related to the system description and instances.

For human activity systems, the System Service provided can be instantiated once or multiple times. For example, an Enterprise defines its financial system and then produces it once for utilization in various parts of the organization. In another example, an Enterprise such as a bank may provide a value added financial advisory system service for its customers. In this case the definition of the advisory system is used as the "blueprint" to establish the service at various branches of the bank. Such systems can also be retired where instances are removed or the entire System-of-Interest is terminated.

Project Scope

The enterprise Change Management function establishes and monitors projects that accomplish structural transformations during the life cycle. In doing so, the project planning must identify the scope, the cost and the schedule to be followed. In regards to scope, it may include a part of a stage, an entire stage or multiple stages within the life cycle. Thus, we can interpret the System Coupling Diagram in Figure 4-5 as meaning the scope of the Situation System treated by the project. For complex Systems-of-Interest the planning of execution of projects is an essential part of the early phases of the life cycle. As was illustrated in Figure 3-6, this can involve the utilization of a Project Design enabling system.

As presented in Chapter 3 in relationship to the utilization of ISO/IEC 15288 projects select and as necessary "tailor" technical processes to be utilized in performing the work activities. We now consider the usage of these processes within the scope of a project.

From Requirements to Architecture

The following ISO/IEC 15288 Technical Processes are utilized in defining a viable system solution:

Stakeholder Requirements Definition	define the requirements for a system that can provide the services needed by users and other stakeholders in a defined environment
Requirements Analysis	transform the stakeholder, requirement-driven view of desired services into a technical view of a required product that could deliver those services
Architectural Design	synthesize a solution that satisfies system requirements

Note: The terms System of Capabilities, System of Requirements and System of Functions/Objects are not part of ISO/IEC 15288. This view of life cycle work products has been developed by your author in order to develop a unified system approach where the System Coupling Diagram provides a generally applicable mental model.

The processes are tailored to meet the specific needs of various types of systems, various development strategies as well as the specific needs of organizations (enterprises) and their projects. They could even be tailored to adapt the perspective of successive system versions that has been portrayed in Figure 4-1.

Various forms of descriptions are produced as outcomes of applying the processes. As noted, with the system perspective provided above, it is useful to view each description as a version of the system. That is, from a System of Capabilities reflecting the need, a requirements document is an early version of the system. As a result of analyzing the requirements, the services to be provided by the system, often described as functions represent another version of the system. The architectural design leads to an identification of system elements and their interrelationships as a viable system design solution. These descriptions provide another version of the system.

The various versions reflect various concerns of stakeholders resulting in a variety of views of the system. A common classification of some useful views is as follows:

capability view – Describes the capabilities that need to be provided in order to achieve the desired system services.

operational view - Describes how the system will be utilized. Various "use cases" are identified that represent the different means of system utilization.

functional view - Describes the functions that will be required in order to provide the required system services.

object view – Describes the objects and their interactions that will be required in order to provide the required system services.

physical view - For defined physical systems, this view describes the structure of the system elements and interrelationships that will be utilized in the realization of the physical system.

activity view – Defines concrete activities of processes or procedures to be carried out in providing a service.

For non-trivial systems, there are typically multiple versions of requirement documents, functional and/or capability descriptions, and alternative architecture solutions that reflect the various views. This is the result of a development strategy involving iteration in the early stages of the life cycle until a satisfactory solution is obtained.

Note: The processes performed during the early "front-end" stages of the life cycle lead to a System Architecture. A companion standard to ISO/IEC 15288 is ISO/IEC 42010 [ISO/IEC 42010, 2010] that is concerned with Architecture Description. Since architecture has such a strong influence on eventual System Products and System Service, some of the important aspects of this standard will be addressed later in the chapter.

Baselines and Configurations

Baselines are used to establish a stable reference point, both within a project and for a system under development. Baselines are also utilized as reference points for the resulting deployable products and/or services. Project baselines track the project parameters of scope, cost, and schedule that were identified earlier. Through this baseline information, project management can evaluate where the project stands and what type of corrective action may be required in order to achieve project goals. Further, the baseline information is used as feedback to the Change Management function for their use in evaluating where the project stands in relationship to the life cycle.

System descriptions of a System-of-Interest that are viewed as being stable are also frozen at various stages of the life cycle in the form of baselines typically called configurations. For example, baseline configurations can be established for versions of capability descriptions, requirement descriptions, functional or object descriptions, physical descriptions, activity descriptions or as a collective description of some or all of them. It is desirable to formalize certain configurations that form a basis for configuration management. In managing the configuration, all further changes must be traceable to the baseline configuration.

Produced Instances

When products and/or services are produced from system descriptions, it can be appropriate to manage configurations of realized instances of a system. The instances provide full or partial services and are managed as products or packaged services provided to acquirers (customers). For some systems, particularly computer-based systems the customer may generate and manage their own configured instances.

Based upon the type of system, the needs of the enterprise and its projects, the following ISO/IEC 15288 Technical Processes are applied in producing instances.

Implementation	produce a specified system element
Integration	assemble a system that is consistent with the architectural design
Verification	confirm that the specified design requirements are fulfilled by the system

These processes are involved in the generation (production) of configurations as well as in their verification in respect to architecture design and stakeholder requirements.

The following ISO/IEC 15288 processes are involved in preparation for, the validation of, the operation of, as well as the maintenance of an instance of a configured system.

Transition	establish a capability to provide services specified by stakeholder requirements in the operational environment
Validation	provide objective evidence that the services provided by a system when in use comply with stakeholders' requirements
Operation	use the system in order to deliver its services
Maintenance	sustain the capability of the system to provide a service

The following ISO/IEC 15288 Project Supporting Process is aimed specifically at the management of baselines as well as configurations.

Configuration Management	establish and maintain the integrity of all identified outputs of a project or process and make them available to concerned parties

In addition to baselines and configurations and based upon the nature of the System-of-Interest, the nature of the enterprise as well as policies and procedures, other forms of version control such as releases, updates, and upgrades may also be used to identify managed system descriptions and/or instances.

Regardless of the classification scheme applied to System(s)-of-Interest, it is essential to provide a means of uniquely identifying the various versions, via policy and procedure at the enterprise level and/or at the project level. Further, the policy and procedures should include directions for building an information model for versions making the information available to interested parties as well as for archiving.

Operational Parameters

Whether a change to operational parameters for a system product or service leads to a new configuration, release, update or some other form of named version is a question that must be related to the nature of the system product or service as well as the enterprise policy and practice. Even though an alteration of parameters involves a change resulting in a modified behavior they are typically not treated as fundamental changes in system descriptions. However, it is vital to have policies and procedures for keeping track of actual parameters over time and the utilization effect results obtained via the various parameter configurations.

ARCHITECTURE DESCRIPTION

As noted above, there is an international standard ISO/IEC 42010 (Architecture Description) that has been developed in order to provide guidance in architecting systems. This standard is based upon a previously successful standard IEEE 1471 that provided guidance for architecting software intensive systems [Maier, Emory and Hilliard, 2004].

The top-level model of the standard provides a good summary of the relationship between systems, stakeholders, architecture and environment as portrayed in Figure 4-6.

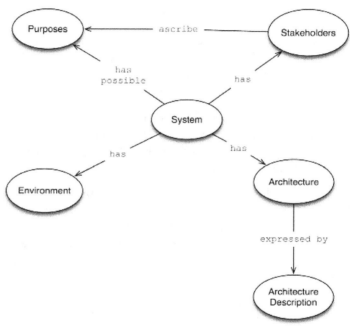

Figure 4-6: Top-Level Architecture Concepts and Principles (reprinted with permission of SIS Förlag AB http://www.sis.se)

The model is completely consistent with the view of a system that has been successively developed in this book. Each one of the elements in the figure identifies concepts that should now be familiar to the reader.

An important aspect of describing System(s)-of-Interest is the differentiation between views and viewpoints. We have introduced these two words in earlier chapters that are now provided with the definition from ISO/IEC 42010.

architecture view - work product expressing the architecture of a system from the perspective of system concerns

architecture viewpoint - work product establishing the conventions for the construction, interpretation and use of architecture views

It can be noted that both are work products where in the case of a viewpoint it relates to the selection of the means of description used in producing views that are related to stakeholder concerns. These aspects of architecture are incorporated in the second level model that defines the concepts and principles of an Architecture Description as portrayed in Figure 4-7.

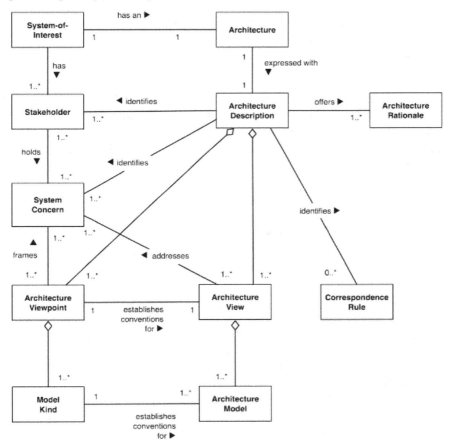

Figure 4-7: Concepts and Principles of Architecture Description (reprinted with permission of SIS Förlag AB http://www.sis.se)

The elements and relationships portrayed in this figure follow the rules of UML (the Unified Modelling Language) [ISO/IEC 19501, 2005]. For those not familiar with this notation, the following guidance is provided.

The notation (x..y) is used to indicate cardinality (that is, number of instances); for example (1..*) indicates at least one but an arbitrary number of additional instances, (0..1) indicates that there can be no instance or at most one instance.

The following verbs or verb phrases that connect the elements use this form of identification to express the cardinality as indicated in the following illustrative principles that can be extracted from the figure:

- The System-of-Interest has 1 Architecture
- The System-of-Interest has at least 1 or multiple Stakeholders
- Stakeholder holds at least one or multiple System Concerns
- ...

The generation of a complete list of principles based upon the concepts provided in the figure is left as an exercise for the reader. In particular note the relationship between viewpoints, views and models that is consistent with the earlier discussion of these system aspects.

An interesting aspect of the standard is related to the Architecture Rationale which is required to justify the selection of architecture features. Another aspect is the possibility of declaring the means of correspondence between various architectural models and the identification of the principles (rules) that are involved in the correspondence.

Architecture Frameworks

Architecture frameworks are defined and described in the ISO/IEC 42010 standard as follows:

architecture framework - conventions, principles and practices of architecture description established within a specific domain of application or community of stakeholders

> "An architecture framework establishes a common practice for creating, interpreting, analyzing and using architecture descriptions within a particular domain of application or stakeholder community. An architecture framework serves as a basis for creating architecture descriptions; a basis for developing architecture modelling tools and architecting methods; and as a basis for processes to facilitate communication, commitments and interoperation across multiple projects and/or organizations."

The establishment of rules and conventions for viewpoints and the views is a valuable unifying factor in for example, managing a product line, organizing a group of related projects or establishing acquirer-supplier relationships for acquisition of system products and/or services. Architecture frameworks have also become popular as a means of collectively describing systems that are or interest to an enterprise; so-called enterprise architectures as described in the final chapter of this book.

A growing number of architecture frameworks have been established with the goal of assisting various communities of stakeholders in describing system architectures. One of the first frameworks was developed by John Zachman and bares his name, the Zachman Framework for Information Systems. [Zachman, 1987 and 2008] As a response to disorderly and costly acquisition activities, the US Department of Defense successively developed a Framework called DoDAF. In the United Kingdom, the MoDAF (Ministry of Defence) framework was created. There also exists a NATO framework called NAF. On the non-military side, there have arisen architecture frameworks for US Federal Enterprises (FEAF) and The Open Group consortium created TOGAF that instead of focusing upon system products, provides a framework for processes. There are several others as well. A web search on Architecture Frameworks or any of these individual frameworks will yield many references.

Many of the frameworks have been developed by committees and consortiums and have become quite complex. One can even question the viability of the frameworks when it takes more time to learn about the framework (several hundred pages of description) than it takes to describe the system architectures that are of interest.

It is also interesting to note that several of the frameworks including DoDAF and MoDAF are now trying to retrofit the guidance provided by the ISO/IEC 15288 and ISO/IEC 42010 standards into their framework approaches. This retrofit will most likely add complexities to the already complex frameworks.

A LIGHT-WEIGHT ARCHITECTURE FRAMEWORK

In face of the growing complexity of architectural frameworks, there have arisen new framework approaches that reduce the complexity. Your author has been developing such an approach to architectural frameworks that builds upon an integration of the concepts and principles established in the ISO/IEC 15288 and ISO/IEC 42010 standards as well as the concrete system semantics that has been presented in the System Survival Kit. Thus, in a manner similar to the balancing aspects of architecture, processes, methods, and tools portrayed in Figure 4-3, a better balance is sought in producing a Light-Weight Architecture Framework (LAF) as portrayed in Figure 4-8.

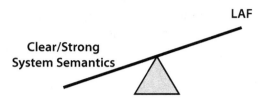

Figure 4-8: Balance System Semantics and Architecture Framework

Your author claims that via the establishment of the limited set of clear and strong system concepts and principles of the System Survival Kit that a Light-Weight Architecture Framework according to the balance shown in Figure 4-8 is feasible and desirable. One can consider the opposite situation where system concepts and principles are not well established or there exist a large number of concepts and principles that become difficult to understand and utilize result in a heavy architecture framework.

LAF, in an ISO/IEC 15288 consistent manner, builds upon the fact that as a system progresses through its life cycle stages the concerns of various stakeholders are made evident. Thus, definition, development, utilization and eventual retirement of systems gives rise to a variety of views held by diverse stakeholders as portrayed at the beginning of this chapter (Figure 4-1).

Thus work products produced by processes that are executed in projects during the life cycle stages are related to concerns and views that describe capabilities, requirements, functions/objects, products, services and lessons learned. The views expressed via viewpoints, correspond to system descriptions representing the togetherness (relationship and connection) properties of elements of the view. For example togetherness can be represented by text, structured text, or models such as system maps showing static groupings of related elements, diagrammatical or pictorial representations of elements and flows of material, data, information and/or energy or control of sequencing in the case of activities carried out by machines or humans. With this perspective, we can constitute that an architecture description is based upon a collection of described systems representing different life cycle related views of a System-of-Interest developed according to viewpoints as enumerated in Table 4-1.

Table 4-1: Stakeholder Views and Potential Viewpoint Description Methodologies

Stakeholder	Views	Potential Viewpoints – Model Kinds
Owner	Capabilities	System Maps, Rich Pictures, Entity-Relationship, Systemigram
Conceiver	Require-ments	Requirements Structured Text, Entity-Relationship, Influnece Diagrams, Use Cases, Systemigram
Developer	Functions/ Objects	Functions/Objects IDEF, Class Diagrams, UML, SysML
Producer	Product/ Service	Product/Service Parts List
User/Main-tainer	Service Delivered	Service Delivered Behavior Diagrams, Use Cases, Entity-Relationship, Systemigram
All	Lessons Learned	Stories, Archetypes, Metrics

There are a variety of modeling methodologies that can be used to produce the various system descriptions as work products as indicated in Table 4-1. The following briefly describes some of the methodologies:

System maps – Systems maps are essentially structure diagrams. Each element or sub-system is contained in a circle or oval and a line is drawn round a group of elements or sub-systems to show that the things outside the line are part of the environment while those inside the line are part of the system. There are NO lines connecting elements, sub-systems or systems in a systems map; it is purely a statement of the structure as you see it in your mind.

Influence diagrams – These are developed from systems maps and indicate where one element in the situation has some influence over another. Arrows indicate the direction of the influence and the lines between elements may be of different thickness, shading or color in order to distinguish strong and weak influence. Strictly speaking, influence should only be shown from elements at a higher or at the same level in the system; that is to say, subsystems cannot influence systems and sub-systems and systems cannot influence the environment.

Entity Relationship Model (ERM) – ERM in software engineering is an abstract and conceptual representation of data. Entity-relationship modeling is a relational schema database modeling method, used to produce a type of conceptual schema or semantic data model of a system, often a relational database, and its requirements in a top-down fashion.

Systemigrams – is a word derived from "systemic" and "diagram"' and portrays a System-of-Interest which is described by text that is structured according to systemic principles. A Systemigram should not be constructed in attempt to capture first thoughts, but rather as a translation of the words and meanings that appertain to a piece of structured writing.

Unified Modeling Language (UML) – UML is a standardized general-purpose modeling language in the field of software engineering. UML includes a set of graphical notation techniques to create abstract models of specific systems.

Systems Modeling Language (SysML) – SysML is a Domain-Specific Modeling language for systems engineering. It supports the specification, analysis, design, verification and validation of a broad range of systems and System-of-Systems. SysML was originally developed by an Open Source specification project, and requires an open source license for distribution and use. SysML is defined as an extension of a subset of the Unified Modeling Language (UML) using UML's profile mechanism.

Integration DEFinition (IDEF) – IDEF is a family of modeling languages in the field of systems and software engineering. They cover a range of uses from function modeling to information, simulation, object-oriented analysis and design and knowledge acquisition. These "definition languages" have become standard modeling techniques.

Class diagrams – In the Unified Modeling Language (UML), a class diagram is a type of static structure diagram that describes the structure of a system by showing the system's classes, their attributes, and the relationships between the classes.

Rich pictures – Rich pictures were developed as part of Peter Checkland's Soft Systems Methodology as an approach to help capture appreciation and understanding of messy complex situations.

The selection of the viewpoint description methodologies and model kinds is central to systems life cycle management. In fact, since the methodologies are systems themselves, they become a part of the Enterprise systems portfolio. That is, they become the standards for predefined views that indicate what type of viewpoint definition is to be deployed for the work products related to processes applied during the various stages. Ideally, the set of viewpoints can become an enterprise standard for all system assets or for classes of system assets as well as Respondent Systems that are to be life cycle managed. As mentioned earlier the framework can support those systems that are relevant for development of a product line or support the parties involved in an acquirer-supplier relationship concerning standards for description and communication. On a broader scale, an architecture framework such as LAF can be used in an organizational or enterprise wide context thus providing standards for description and communication to all enterprise stakeholders. This use of frameworks will be discussed in the final chapter of the book when organizations and their enterprises are examined as systems.

OWNERSHIP OF SYSTEM DESCRIPTIONS AND INSTANCES

In many larger organizations, there is a tradition of system ownership vested in individuals or in some cases enterprise entities (groups or teams). Ownership implies authority and responsibility to create, manage, (perhaps operate) and dispose of a System-of-Interest.

For institutionalized infrastructure Systems-of-Interest that are entirely owned by an enterprise or parties thereof, the entire life cycle management responsibility including operation is vested with these system description owners. For example, system assets as portrayed in Figure 1-6 and illustrated in Table 1-1 as well tailored life cycle models and sets of system management processes based upon ISO/IEC 15288. These systems belong to the system asset portfolio of an enterprise and provide the system resources from which Respondent Systems are configured in handling Situation Systems including the planned systems that are developed during life cycle management.

Trading in System Products and Services

For instances of systems that are acquired by an enterprise for utilization (operation) as an asset, the system description is owned by the supplying organization. In such cases, the acquirer owns or licenses an instance of the system as a product or service for which rights of utilization are provided via an explicit or implicit acquirer-supplier agreement (in some cases in the form of a contract). To illustrate this acquirer-supplier relationship, consider Figure 4-9.

The figure indicates how a supplying enterprise from a system description template generates instances of the product or service for delivery to an acquiring enterprise. The acquiring enterprise then utilizes the system instance as an asset in achieving its purpose, goals and missions.

Naturally, the supplying enterprise has infrastructure system assets in their own portfolio that assist them in achieving their purpose, missions and goals of product and/or service development, production and support.

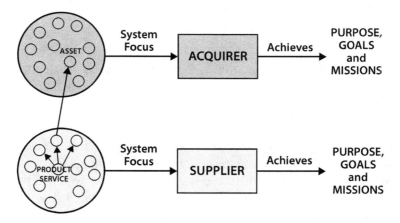

Figure 4-9: Trading in and Ownership of Systems

The enterprises are involved in making decisions related to the life cycle management of the system description, respective instances. In cases where the supplier provides a unique product and/or service to the acquiring organization, an agreement can be established in which both enterprises participate in the ownership and life cycle management of the system description. Whereas, for mass-produced products supplied to a variety of acquirers (consumers), the ownership of the system description typically remains with the supplying organization.

The ISO/IEC 15288 Agreement Processes are provided to assist in structuring relationships between acquirers and suppliers in a supply chain. The two processes are as follows:

Acquisition	obtain a product or service in accordance with the acquirer's requirements
Supply	provide an acquirer with a product or service that meets agreed requirements

The processes are to be used for enterprise external as well as internal acquisition and supply relationships. The outcomes of the processes as well as the activities are tailored in order to meet the specific needs of the enterprise and its agreements.

Supply Chain Relationships

Figure 4-9 portrays one link in a supply chain. The supplier in turn can well be an acquirer of system products and services delivered by other suppliers that are assets in their system portfolio and used in the process of providing their products and/or services. For complex systems the supply chain for suppliers providing products and services can be rather long. Some examples include the supply chains related to various life cycle stages of an automobile System-of-Interest or health care information System-of-Interest. In such cases the products and services delivered by suppliers may become part of the life cycle management of a system description owned by the supplier. These outsourcing relationships and their impact upon the three fundamental changes of systems are portrayed in Figure 4-10.

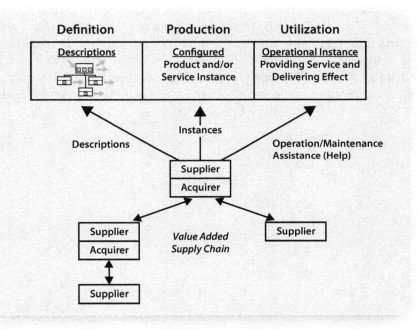

Figure 4-10: Outsourcing Relationships (Supply Chain)

In this figure, it is important to observe in respect to Definition that in an agreement with a supplier the outsourcing can involve delivering complete system description solutions or portions thereof. For example, a supplier could, given a system of stakeholder requirements developed by the acquirer, develop and supply a system for the functional or object based architectural design solution. The supplier in turn can be an acquirer of portions of their delivered results by outsourcing to other suppliers.

In respect to Production, the outsourcing agreement with a supplier can vary from total production responsibility to the supply of instances of system elements to be integrated by the acquirer. Once again these suppliers can be acquirers of portions of their delivery from outsourcing to other suppliers.

In respect to Utilization, for non-trivial systems, outsourcing agreements can be made with a supplier to provide for operational services, for example, operating a health care information system. Further agreements with suppliers can involve various forms of logistics aimed at sustaining a system product or service or for supplying assistance in the form of help desks. Once again suppliers that agree to provide services related to utilization can be acquirers of the services of other suppliers.

Important to all supply chains is that supplying parties contribute some form of added value to the life cycle of a System-of-Interest. The proper management of a supply chain system asset is a vital part of the operations of an enterprise. In fact, the supply chain itself is a System-of-Interest that is composed of acquirers and suppliers as system elements. There definitely is a structure tied together by agreement relationships. Further, the operation of the supply chain definitely results in an emergent behavior. The supply chain system becomes a vital infrastructure asset in the system portfolios of enterprises and forms the basis for extended enterprises.

KNOWLEDGE VERIFICATION

1. Explain the difference between the system description of a System-of-Interest and instances of the system.

2. Provide examples of various defined physical systems as well as defined abstract and human activity systems as well as systems composed of mixture of the three categories.

3. Consider some systems with which you are familiar and, if possible, identify the driving concepts and principles upon which the systems are defined.

4. Discuss the affects of not having a strong architecture and not having well defined processes in developing and implementing Systems-of-Interest.

5. Explain how the System Coupling Diagram can be applied for capturing work products as the end state of Situation Systems during various stages of a System-of-Interest life cycle.

6. Explain the role of a capabilities view, operational view, functional view, object view, physical view and activity view in the description of a System-of-Interest.

7. What are baselines and configurations and how are they related to the life cycle of a system-of-interest?

8. Describe how versions of system baselines and/or configurations are identified for Systems-of-Interest that are owned or operated by an enterprise with which you are familiar.

9. Compare the concepts and principles used in the Light-Weight Architectural Framework with other architectural frameworks.

10. Identify Systems-of-Interest that are completely owned and operated by an enterprise with which you are familiar.

11. Identify systems or portions thereof that are acquired by an enterprise with which you are familiar.

12. Describe outsourcing relationships in respect to Definition, Production and Utilization of a System-of-Interest with which you are familiar.

13. Given that a supply chain can be viewed as a System-of-Interest for an enterprise where acquirers and suppliers are system elements, describe how relationships between the elements are defined.

Interlude 3: Case Study in Architectural Concepts and Principles

In Chapter 4, one of the system success factors described is the importance of identifying and utilizing driving concepts and principles in achieving organized simplicity. In order to demonstrate this importance in a real situation in which your author was active as architect, we consider the development of the worlds first microprocessor controlled Automatic Train Control System. This system was developed for the Swedish Railways (SJ) in the latter part of the 1970s. The case study has been extracted from two published papers describing this system; namely [Lawson, et.al., 2001] and [Lawson, 2008].

INTRODUCTION

Railways as we know them today had their origin in the United Kingdom with the first public railway in 1825. At that time, there were 25 miles of track and 2 locomotives. In 1829 Stevenson's steam engine the Rocket was introduced and in competition with other engines attained a speed of 29 mph (unloaded) and 25 mph hauling 13 tons of wagons. This was the catalyst that led to the rapid development of railroads around the world. By 1875 there were approximately 160,000 miles of track and 70,000 locomotives in the world. This is an astoundingly rapid development especially considering the primitive means of international transportation and communication available at that time. It is interesting to compare this with the rapid expansion of automotive traffic as well as computing technology and the Internet.

Early accidents due to human errors in the UK and elsewhere rapidly led to the development of signaling to control traffic. To provide this critical function, several mechanical interlocking solutions where developed in order to prevent signalmen from accidentally setting conflicting routes. Interlocking developments then proceeded through generations of an ingenious variety of more complex mechanical and electromechanical systems.

Today, the safety of millions of train passengers is dependent upon reliable safety related equipment and functions in the entire railway system. One of the most important functions is the monitoring of the behavior of train drivers; that is, assuring that they abide by speed limits, signal status and other conditions. There have been numerous train accidents in Europe and elsewhere in the past where the availability and proper operation of this function would have hindered these incidents. This function, now often referred to as Automatic Train Protection (ATP), has been implemented since 1980 in Sweden as the Automatic Train Control (ATC) system.

AUTOMATIC TRAIN CONTROL IN SWEDEN

The availability of inexpensive microprocessors and electronics in the mid-1970s offered new solution possibilities for interlocking as well as for protecting against driver errors. The Swedish National Railways (SJ) was quick to exploit these new possibilities and ordered the development ofv the worlds first computer-based interlocking and speed control system. The investment in this solution was motivated as follows:

- -To meet demands of increased efficiency of railway transportation on both existing and new tracks, the train speed must be increased and the trains must operate with shorter intervals.
- This requirement increases the demands on both the safety system and the train drivers thus leaving little room for human errors.
- The high degree of accuracy of the ATC system minimizes the risks for driver error.

Initially (in 1980 when the first ATC systems where installed), the plan for the Swedish state railways (SJ) was that the train should be driven entirely according to the external optical signals, and that the ATC system should be considered only as a safety back up. With the introduction of the X2000 high-speed trains (200 km/h) in the early 1990s, it turned out that the optical system was insufficient for presentation of all information needed, e.g. earlier warning for restrictions ahead, and different speeds for various train types. Also, after operational experience with the ATC system had been accumulated, it turned out that the ATC system could be trusted for presentation of information not otherwise available along the track. The resulting system nowadays is a very efficient, robust, and safe combination, well matching more expensive and more complicated systems being used elsewhere in the world.

If the driver should lose concentration for a moment, the ATC system will then take over the control of the train by applying the brakes. This brake application continues until the driver manually acknowledges to the system that he is once more capable of controlling the train. If the driver should fail to regain control, the ATC will continue to brake the train to a standstill.

The two major technical functional constituents of the ATC system are the track to train transmission system product and the on-board system product.

The Track to Train Transmission System

The wayside equipment consists of track-mounted transponders (called balises) transmitting messages (telegrams) to the vehicle when activated by the antenna mounted on the vehicle (see Figure 1). The information transmitted includes signal status as well as the speed limit to be followed until the next transponder group. Each type of information generates a unique message (telegram). The transponders are combined into groups of minimum two and maximum five transponders. A transponder group can be valid for the current or the opposite direction of travel, or for both travel directions.

The transponders in a group can have either a fixed code or be coded by an encoder connected between the signaling system and the transponder, in such a way that the transponder group can give information corresponding to the current signal aspect to the on-board equipment.

When a vehicle with an active ATC travels over a transponder group, each transponder will be activated by the energy received from the antenna of the vehicle. The coded message is continuously transmitted to the vehicle equipment as long as the transponder is active. A valid combination of transponders will transmit all the information necessary for the vehicle equipment to evaluate the message and take the required action. The on-board equipment will detect either a faulty message or an invalid combination of transponders and notify the driver accordingly.

Figure 1. ATC Track to Train Transmission System

The On-Board System

The vehicle on-board equipment is portrayed in Figure 2 and consists of the following major components:

- An antenna mounted underneath the vehicle that activates the track equipment (transponders) by continuously transmitting a powering signal and receiving transponder messages to be evaluated and used to supervise the safe travel of the train.
- A set of computer equipment that evaluates the transponder messages, presenting the information to the driver and braking the train to a safe speed level if the driver should fail to take the correct actions, i.e. not braking the train or exceeding speed limits. The driver has to manually cancel each ATC brake application by pushing a brake release button.
- Cab equipment consisting of a driver's ATC panel used by the driver to enter data that is relevant to that specific train, and all other communication with the ATC equipment. The panel also keeps the driver informed of current speed limits and target speed limits at speedboards and signals ahead.
- Vehicle interfacing devices, such as speedometer connection, main brake pipe pressure sensor and one or more brake valves.

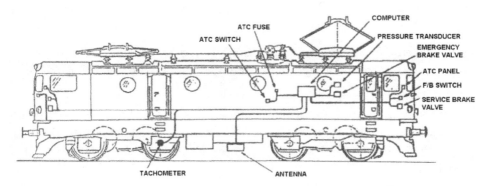

Figure 2. On-Board ATC System

In order to provide for fault-tolerance, a three-processor solution with majority logic comparison of outputs was utilized for the early versions of the on-board system. Due to the observed high reliability of the hardware based upon many years of operation, later versions of the on-board system only utilize two processors.

Ownership of the System Products

The track to train transmission product was developed and delivered by Ericsson Signal AB (now owned by Bombardier). Two versions of the on-board product were developed; one by Ericsson Signal AB for the Stockholm Local Trains and one version for the Swedish Railways (SJ) by Standard Radio AB (at that time owned by ITT). The Standard Radio system has since 1990 been owned by ATSS (Ansaldo Transporti Signal System). Your author was the architect of the Standard Radio system that is used in the majority of the locomotives in Sweden and which has continually been further developed and supported by Teknogram AB of Hedemora, Sweden. [www.teknogram.se]

EVOLUTION OF THE ARCHITECTURAL CONCEPTS

In 1975, your authors consulting services were contracted by Standard Radio to assist the chief designer Sivert Wallin in the conceptualization of the architecture. Following a review of the work done to date on the software, Harold Lawson and Sivert Wallin re-examined the fundamental requirements of the ATC function and developed the problem oriented architecture concepts that has successfully provided product stability as well as a sound basis for further development under the entire life cycle of this ATC on-board system product. The following three core concepts were developed and have been driving factors during the product life cycle.

Time Driven -The major conceptual aspect of the design is the treatment of the system as being continuous in time as opposed to being discrete event driven.

> Motivation - Given the fact that a 250 millisecond resolution (dT) of the state of the train in respect to its environment was determined to be sufficient to maintain stability, it became clear that the simplest approach was to simply execute all relevant processes (procedures) during this period of time.

Software Circuits(*) – As the result of the time driven concept a cyclic time driven approach became the basis for the solution where short well-defined software procedures behave like circuits.

(*) The naming of this concept was developed later when the concepts of the architecture were re-applied in a Swedish research and development project for local area networks in vehicles [Hansson, et.al, 1996] and [Hansson, et.al, 1997].

Black-Board Memory – In order for Software Circuits to have access to key information, variables are retained in a black-board where both reading and writing are permitted.

This simplification of concepts led to the fact that the processors only needed to be interrupted by two events. One interrupt to keep track of time (1 millisecond) and one interrupt when information from a transponder is available. The 250 milliscond dT is more than adequate to perform all processing. Adding more structure to the problem, for example, via the use of an event driven operating system approach would have had negative consequences in terms of complexity, cost as well as reliability and risk thus affecting safety. This simple operating system organization is illustrated in Figure 3.

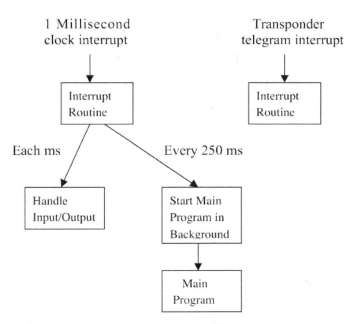

Figure 3. ATC on-board Operating System Structure

The "circuit like" structure of the software led to highly simplified coding of processes (procedures). While it would have been useful to deploy a higher-level language in the solution, it was deemed unnecessary due to the low volume of code that was expected. Experience has indicated that this was a reasonable decision at that time. On the other hand, it was decided to comment the code in a higher-level language. In earlier versions of the product, the Motorola MPL (a PL/I derivative) was employed. In later versions, a more Pascal like annotation has been consistently employed. In system tests, MPL, respectively Pascal versions have been executed in parallel with the execution of the assembly language version in order to achieve system verification.

ATC Software Statistics

There have been two major versions developed and two minor variations on the second version that have been developed for on-board utilization by the Swedish Railways (SJ). The size, in terms of number of procedures, lines of assembly code and number of memory bytes is indicated in the following table.

Version	Number of Procedures	Number of Instructions	Number of Bytes
ATC1	157	4116	10365*
ATC2	308	10281	26284**
ATC2.1	313	10523	27029**
ATC2.2	339	11178	29522**

* Motorola 6800 microprocessors
** Motorola 68HC11 microprocessors

The small size, clear structure, and simplicity of the software solution have led to many advantages in respect to verification as well as in further development and maintainability. It is interesting to note that the first version ATC1 was in operation from 1980 until 1993 and during these fourteen years of operation not a single line of program code was altered. Perhaps a world's record!!!

The ATC2 version was developed to provide support for the X2000 high-speed trains. New functionality was added but the architectural concepts where not changed.

The two latter developments of the system led to ATC2.1 where a radio based control instead of the train to track balise transponder system was employed. Further ATC2.2 was developed for integration in the X2000 train sets and other passenger and freight trains that travel over the Öresund bridge between Sweden and Denmark. In this case, Teknogram also developed an interface PC-board and software based upon the same operating system as ATC2 for communication with the Siemens solution utilized on the Danish railways. This system began operation during the summer of 2000 when the bridge officially opened. Now, even the line between Helsingor in Denmark and Helsingborg in Sweden also deploy this dual solution. The different software versions are fully backwards compatible, i.e. ATC2.2 could be used in any train in Sweden and Norway if desired.

In addition to the main ATC on-board product, a separate PC-board and software running under the same operating system solution was developed to function as the "black box" recorder for ATC. The recorder collects information for up to three days of train operation and includes telegram information and all transitions of speed greater than 2 km per hour. The most recent version of the recorder utilizes flash memories. Earlier versions utilized solid-state memories that required constant power (battery back-up).

It was hoped by Standard Radio that ATC would be a successful export product. Unfortunately, this market did not fully materialize until later. Only a small project in Perth, Australia, was implemented with ATC1 (and is still operating and expanding). Several potential customers, including British Railways examined the product, but decided not to buy it. This is very unfortunate since it has now been proved that it has worked reliably for train traffic for over 30 years. This is a truly impressive record. The cost of one single serious accident would most likely pay for the installation of the system not to mention the personal loss and suffering associated with such accidents.

Since 1990 the solutions utilized in ATC1 and ATC2 have been further exploited by Ansaldo (ASTS). ASTS along with Teknogram have been involved in several installations of ATC. The installations have included an ATP (Automatic Train Protection) system for Keretapi Tanah Melayu Berhad of Malaysia (installation 1996), ATP for Hammersley Iron Ore Railways in Australia (installation 1998), the ATC system for Roslagsbanan in suburban Stockholm (installation during 2000), ASES (Advanced Speed Enforcement System) for New Jersey Transit in USA and the monorail system for Kuala Lumpur, Malaysia. All of these on-board systems have been based upon the same architecture and operating system core solution. However, the programs for the latter solutions are written in the Ada programming language.

Further, Teknogram AB has successfully utilized the same architecture and operating system to develop and market more than 20 train simulators. Consequently, this on-board ATC architecture has been the basis for the Teknogram business concept. For Teknogram and Ansaldo, this represents a truly exceptional example of the reuse of architecture concepts and operating system core for the implementation of new system products.

Life Cycle Implications

As the concepts evolved, the more global implications of the concepts became evident as documented in a comprehensive software plan presented by your author in 1976.

> *"A comprehensive plan for the specification, development, testing, verification, production and maintenance of the software components of the ATC project is presented. The goal is to produce reliable software parts to complement the three processor Motorola 6800 system so that a trustworthy total system is provided. A further goal is to assure that the software constituent remains reliable under the lifetime of the product. That is, that future modifications to the software will not affect the reliability due to oversights concerning design features and software component interrelationships."*

> ...

> *"The key to a successful software product lies in the ability to decompose the system to be implemented into well defined units such as processes,*

procedures, blocks, etc. Further, the operation, inputs, and outputs of these units must be well specified and the specification must serve as a control over the implementation, testing, production, and maintenance."

...

"In the ATC project, the process is the unit to which the system structure has been decomposed. A process should be viewed as a testable component, precisely as a hardware component (integrated circuit). It must have a clear specification and have a well defined component test procedures."

...

"A system can never be more reliable than its components and their inter-connections. Assuming that each software component has been tested, the interconnections of subsystems of components and finally the total system must be developed, tested, and verified systematically."

Thus, it is clear that even at this early point in the product history conceptualization, the importance of architecture as a controlling factor for the life cycle of the product was clearly identified. Even though the owners of the product and development and maintenance has changed management, the fundamental concepts established in the mid-1970s are still in place and have led to a successful solution for train safety not only in Sweden, but in other countries.

DEVELOPMENT AND
MAINTENANCE PRINCIPLES

The early development work was based upon using a PDP-15 computer both for simulation as well as for assembly language translation. The target system based upon Motorola 6800 processors was connected to the PDP-15 so that both procedure and system testing could be well controlled.

Due to the simplicity of the architecture, many advantages were discovered and principles that guided both development and maintenance where established as follows:

- The structure of procedures provided clear points of built-in controls that aided in testing and fault isolation.
- The stack pointer must be returned to the same point in each execution cycle providing a general control of proper cycle execution.
- No wild loops can occur.
- No backward jumps are permitted other than in well controlled loops in procedures.

- Quick reliable changes can be made and verified thus reducing costs.
- The operating system core can easily be reused by removing procedures and incorporating new procedures for new functionality (recorder, simulator).

Following these principles has led both a reliable and stable on-board system product as well as a basis for the reuse of code.

LESSONS LEARNED

There are several lessons that can be learned from the Standard Radio ATC on-board system product experience. These lessons could well be applied in other products, particularly safety critical computer-based systems. Some of the most significant lessons are as follows:

Architecture is a key aspect

The definition and consequent deployment of a problem relevant architecture is a key factor for success. While it is important to have well defined work processes for all life cycle stages of a product, a good architecture reduces the need for heavy processes with multiple activities and tasks. Decision-making is simplified when decisions are bounded by the architectural concepts.

Engineering view is superior to software view

Instead of creating significant quantities of software, an engineering view of the functions to be performed was taken. The analogy between hardware circuits and the logic of the software, later identified as software circuits provides a strong, simplifying solution. We can conclude that software, especially in large quantities, is dangerous but can be controlled with the proper engineering viewpoint.

Do not add more structure than necessary

Adding more structure to a solution than necessary for achieving desired behaviors leads to unnecessary complexity thus costs and risks. This pitfall is very common, even for safety critical systems. Operating systems and programming languages that provide elaborate structures for interrupt handling, multi-tasking, etc. complicate verification, further development, and especially maintenance. In addition, complex methods and tools are often deployed. All of these supporting methods and tools implicitly become a part of the product. Together they often are an overkill solution leading to increased cost and risk.

Verification is a vital aspect of safety critical systems

All safety critical systems must be verified in respect to their specifications and safe behavior in various situations. The combination of module testing, code inspection, and system test via simulation has proved to be an adequate approach for ATC. Simplicity in the architecture and code structure simplifies verification and contributes significantly to safety verification.

A good technical solution is essential but does not in and of itself guarantee safety

The technical solution is only one component of the total system. There are many other factors, including investment decisions, human factors, operation management, and so on, that can and have affected the utilization of the ATC safety system.

ACKNOWLEDGEMENTS

Several people have had important roles related to ATC and in particular the on-board system originally developed by Standard Radio. In this regard, your author gratefully acknowledges the contributions of the following people.

Bengt Sterner at SJ/Banverket for his pioneering vision in developing the combined track to train transmission and the on-board microprocessor controlled ATC systems.

Sivert Wallin for his pioneering work at Standard Radio in developing the first on-board system. Founder and president of Teknogram AB, Hedemora, Sweden.

Johann F. Lindeberg and Øystein Skogstad of Norways Technical University for providing insights in the programming of ATC.

Berit Bryntse and others at Teknogram for their continued further development of the on-board system products.

Bertil Friman now employed at Ansaldo Sweden for his work in developing the verification strategy for ATC2.

CASE STUDY REFERENCES

Hansson, H., Lawson, H., Strömberg, M., and Larsson, S. (1995) BASEMENT: A Distributed Real-Time Architecture for Vehicle Applications, Proceedings of the IEEE Real-Time Applications Symposium, Chicago, IL. Also appearing in Real Time Systems, The International Journal of Time-Critical Computing Systems, Vol. 11. No. 3, November, 1996.

Hansson, H., Lawson, H., Bridal, O., Ericsson, C., Larsson, S., Lön, H., and Strömberg, M, (1996) BASEMENT: An Architecture and Methodology for Distributed Automotive Real-Time Systems, IEEE Transactions on Computers, Vol. 46. No. 9

Lawson, H. Wallin, S., Bryntse, B., and Friman, B. (2001) Twenty Years of Safe Train Control in Sweden, Proceedings of the International Symposium and Workshop on Systems Engineering of Computer Based Systems, Washington, DC.

Lawson, H. (2008) Provisioning of Safe Train Control in Nordic Countries, Keynote address appearing in the Proceedings of HiNC2, History of Nordic Computing.

Chapter 5
Change Management

Decisions, decisions ... the only thing constant is change

In striving to achieve purpose, goals, and missions, an enterprise must continually evaluate their problems, opportunities and system(s) situation and take actions to make appropriate operational and/or structural changes. Important questions of authority and responsibility for change, what initiates change, how change decisions are made, who makes decisions, when decisions are to be made, how changes are followed-up are all critical to the enterprise.

There are many theories and significant practice reported in the management literature that is related to decision-making and the treatment of change management. The spectrum of decision-making styles ranges from virtually dictatorial to a universal democracy (everybody involved) in the process of decision-making. In this chapter, critical issues related to decision making and change management are considered and related to the paradigm for thinking and acting in terms of systems as well as the essential aspects of system descriptions and their instances as described in Chapter 4.

ORGANIZATIONAL CYBERNETICS

The field of cybernetics was developed in the mid to late 1940s by Warren Mc-Culloch and Norbert Weiner as a discipline independent means of explaining complex system interrelationships with regard to control, information, measurement and logic. A generic cybernetic system composed of three system elements and their relationships is portrayed in Figure 5-1.

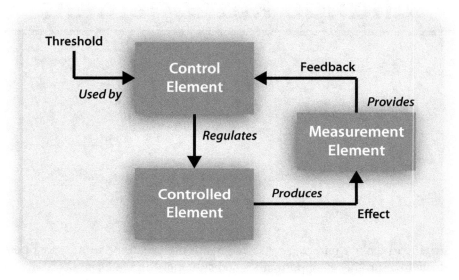

Figure 5-1: A Generic Cybernetic System

A Control Element *regulates* a Controlled Element. The Controlled Element *produces* Effect. A Measurement Element measures the effect and *provides* Feedback to the Control Element. The Control Element compares the current effect with a Threshold value that it *uses* in deciding upon further regulation of the Controlled Element.

Cybernetics is applied in physical systems, for example in the regulation of room temperature. In this case, physical regulation sensors are used to measure the current effect which is fed back to a Control Element where the current effect is compared with a threshold setting which can result in activating or deactivating a heating or cooling element. Such regulation operates continuously as long as the Controlled, Control and Measurement Elements are operational.

Once again, we reinstitute the System Coupling Diagram in Figure 5-2. It is useful to note that some form of control element is necessary for Respondent Systems in which case this element controls the other elements as the Respondent System handles the Situation System. Thus, the cybernetic model is directly applicable whether we are regulating room temperatures, the progression of stages in a life cycle, or a terrorist action. Ashby [Ashby, 1964] in his *law of requisite variety* made the important observation that: Control can be obtained only if the variety of the controller is at least as great as the variety of the situation to be controlled. Thus it is vital that the control element of respondent systems is properly designed and developed so that it is prepared to deal with the complexities of the situation. In fact, control elements should be viewed as system assets that are life cycle managed and instantiated as operative controllers.

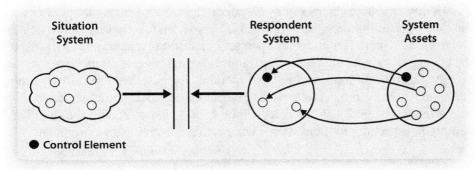

Figure 5-2: System Coupling Diagram Relationship to Cybernetics

While the nature of the control, controlled and measurement elements are different, the principles of cybernetics can be equally applied to non-physical systems. While many others have exploited this similarity, it is Stafford Beer that formalized the usage of organizational cybernetics in what he called a Viable System Model (VSM) [Beer, 1985]. The VSM stipulates rules whereby an organization is, "survival worthy"- that is, it is regulated, learns, adapts, and evolves. Such a learning organization, according to Beer, is constructed around five main management functions; namely operations, coordination, control, intelligence and policy. All of these functions are treated in a cybernetic manner in the VSM. Roughly speaking, the operations function is the Controlled Element. The control function corresponds to the Control Element, intelligence is the result of Measurement that is feedback to the Control Element, policy is used in evaluating a threshold; and finally, coordination deals with the interrelationships between concurrently operative Cybernetic Systems.

Beer also realizes that in complex organizations, several levels of VSM exist and that interrelationships can be described, as systems, via recursive decomposition as described in Chapter 1. In this case Controlled Elements at one level include Control Elements at the next level. Thus, in applying the knowledge related to system descriptions and instances from Chapter 4, observe that from the template system description of the generic cybernetic system instances are instantiated at various levels in the organizational structure.

CHANGE MANAGEMENT AS A CYBERNETIC SYSTEM

The implementation of the Change Management Model presented in the earlier chapters is based upon the principles of organizational cybernetics as summarized in Figure 5-3.

In implementing the model as a system, the Control Element operates continuously according the Observe, Orient, Decide, and Act paradigm. When the Controlled Element is a project it operates in a discrete manner; that is, it makes the change within planned time constraints. For line organizations there may not be specific points of termination. For structural changes, the project form is most common where the project follows the Plan, Do, Check, Act paradigm. Even a line organization charged with the responsibility for a change should use the PDCA paradigm, especially for non-trivial changes to an operational environment.

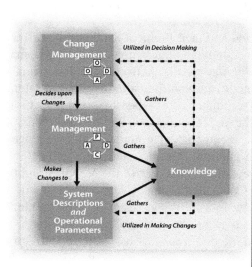

Change Management functions as the Control Element. It uses Knowledge as Feedback in making decisions. Purpose, Goals, and Missions to be accomplished along with policy, rules and regulations serve as Threshold values.

When changes are made via Projects, the Project becomes the Controlled Element. When operational parameter changes are made by a line organization, it is the Controlled Element.

Outputs produced by the execution of ISO/IEC 15288 Technical Processes produce Effect in respect to changes in System Description and Operational Parameters.

The Effect (outputs) produced by all of the elements on the left are gathered as data, interpreted as relevant information and when related to other information becomes the Knowledge that is fed back and fed forward.

Figure 5-3: Change Management System

In contrast to the generic Cybernetic System Model of Figure 5-1, all system elements of the Change Management System gather data that can then be interpreted according to information classification thus providing information. The data can also be measured according to additional relevant information classification and provide more data and information. This corresponds to the Measurement Element of Figure 5-1. By relating information, knowledge is obtained that can be used both as cybernetic feedback as well as fed forward in the form of "know how" capabilities for the actual change related activities. These aspects of assimilating and using knowledge are described in further detail in Chapter 7.

MEASURING EFFECT

Based upon the type of system that is being controlled, various forms of measurements are made in order to determine if the Effect produced by operation of the system meets the threshold of change requirements. The following two measures can be applied to any type of man-made system.

Measure of Effectiveness (MOE)

Measurement of the ability of a system to meet needs expressed as stakeholder requirements. The requirements indicate what the system should be capable of achieving. When measuring the MOE of a system element, the MOE can only be evaluated by determining how well the system element has assisted in meeting the needs of a system-of-interest of which it is a part, meet the system stakeholder requirements.

MOE measurements vary in respect to the type of system involved. For defined physical systems quantitative requirement related measurements can be made. For example, that a climate conditioning system shall maintain a room temperature of no less than 18 and no more than 22 degrees Celsius. For defined abstract and human activity systems, the MOE can involve both quantitative as well as qualitative measurements. For example, an enterprise operates within the limits of an allocated budget (quantitative) as well as meets a requirement of sustaining employee satisfaction (only measurable in a qualitative manner).

Measure of Performance (MOP)

Measures the actual performance achieved due to the inherent system design. Field-testing and/or trials of systems resulting in measurements that can be assessed against some form of performance baseline can determine MOPs.

Quantitative MOP measurements can clearly be applied for defined physical systems; for example that a climate conditioning system due to its design actually sustains an average room temperature of 20 degrees ± 0.5 degree Celsius at least 95% of the time during each day of operation. In the case of defined abstract and human activity systems such as an enterprise, quantitative measurements can also be made in respect to meeting project plans within budget limits. In evaluating performance of qualitative aspects of a system property, like employee satisfaction, measurements can be made by questionnaire or interviews resulting in some form of judgment of the degree of satisfaction.

In summary, while MOE and MOP measures have traditionally been applied to physical systems, it is certainly possible to find corresponding measures for other man-made systems; both quantitative and qualitative. Both measures are essential contributions to the knowledge used as feedback to the Change Management controlling element as well as in providing feed forward guidance for actual change activities by projects or by a line organization.

Customer Satisfaction Index (CSI)

A non-technical "Effect" measurement that is of vital importance for enterprises is the degree of their customer's satisfaction. From a quality point of view, it is customer satisfaction that is the desired effect of a quality management system as stipulated in the ISO 9001 standard [ISO 9001, 2008].

The Customer Satisfaction Index was developed by Professor Claes Fornell [Fornell, 2001] [www.theasci.org]. It is used as predictor of consumer spending and corporate earnings. The CSI model is a set of causal equations that link customer expectations, perceived quality, and perceived value to customer satisfaction. These measurements are in turn linked to consequences defined by customer complaints and customer loyalty – measured by price tolerance and customer retention.

The model has been applied to a wide variety of commercial products and services. A variant of the model provides measurement of satisfaction with the services provided by governmental agencies. The index relates to an initial baseline established from the first year in which measurements where made. Thus, the index shows the successive customer satisfaction increase or decrease relative to that point. The CSI for many industries are frequently published in prominent financial newspapers since they provide an indication of the value of a vital asset, namely satisfied customers.

An important side effect of customer satisfaction that has been observed in making measurements is the degree of employee satisfaction that shows a high degree of correlation with the CSI.

Process Assessment

Another type of "Effect" measurement that can be applied in enterprises is the assessment of process capability. There are a number of Capability Maturity Models that are aimed at measuring human activity capabilities in several areas including software development, system development, acquisition, and others. Based upon the processes that are implemented, the enterprise is evaluated as to how well it executes the processes. Such measurements are based upon the actual achievement of process outputs and are graded in a scale such as the following:

- Level 1 Performed
- Level 2 Repeatable
- Level 3 Defined
- Level 4 Managed
- Level 5 Optimizing

At the initial level, processes are performed and provide basic output results, but are not necessarily repeatable, not well defined, not well managed, and certainly not optimized in order to provide for continuous enterprise improvement. At the higher levels, the enterprise provides assessment evidence that indicates process repeatability, well-defined processes, well-managed processes, and finally that enterprise is continually improving (i.e. optimizing) its processes.

In an assessment, the enterprise is scored for each process and then collectively for the set of processes that are assessed. By binding the assessment to achieved work product outcomes from process execution, the assessment measurements attempt to be quantitative. However, due to subjectivity some measurements tend to become qualitative in nature. In any event, the assessments are valuable inputs for feedback to the Change Management control element as well as feed forward knowledge related to the accomplishment of change.

Note: in the area of system life cycle management addressed by the ISO/IEC 15288, the same ISO/IEC JTC1 sub-committee that developed the 15288 standard has developed a process assessment model for evaluating the processes of the 2002 version of the standard; namely ISO/IEC 15504-6 [ISO/IEC 15504, 2004].

Balanced Scorecards

Balanced Scorecards (BSC) is a control mechanism for enterprises or projects that promote capturing, communication and review of various activities or properties of the top-level system (the enterprise and its projects) in relation to the enterprises or project's strategy and goals. See [Kaplan and Norton, 1996].

A BSC presents various perspectives on the enterprise, typically including:

- the financial perspective,
- the customer perspective,
- the business processes perspective,
- the innovation and learning perspective.

Further perspectives can be added where appropriate. Within each perspective, a set of measurable (or ratable) Key Performance Indicators (KPI's) is defined, which are suitable to demonstrate the current business performance in relation to target values (thresholds). The result is an overview on the strengths and weaknesses on the top level, and can guide the management's overall decision-making process to focus on the most relevant areas for improvement.

A major strength of BSCs is the holistic view on the business it provides as it shows key performance indicators of quite different domains in one overall representation. Consider as an example for an issue stretching over several domains and perspectives: Employees perceive a lack of training with a new IT system (learning and innovation perspective), resulting in sluggish, error-prone customer support services (processes perspective), thus decreasing the customer satisfaction (customer perspective) and slowly lowering sales (financial perspective). In this example, a BSC can serve as discussion basis about actions to be taken, such as further increasing training costs, further IT improvements or the like.

While Balanced Scorecards are useful in obtaining an overview on a variety of topics, subsystems, organizational qualities etc. by Key Performance Indicators, they do not inherently show the interdependencies between these. Systems Thinking methods, as introduced in Chapter 2, such as systemigrams and archetypes are suitable to visualize these interdependencies. E.g. archetypes can be used to point out limiting loops and alternative concepts, which reduce undesirable limiting effects or add new growth loops.

In relation to BSC, Systems Thinking methods can be used to demonstrate;

- the interdependencies between key performance indicators,
- optimizations within a system, which generates a single key performance indicator,
- the relationships of a key performance indicator system to its environment.

Concerning the latter point it is notable, that it can include environmental systems, which are outside the scope of the BSC. These environmental systems can be factors outside the organization (e.g. raw material prices) or within, just not captured by the currently used BSC.

Thus, in implementing a Balanced Scorecard for both the establishment of thresholds as well as a means of measurement in a cybernetic model, the need to measure individually and collectively can well lead to a hierarchy of interacting cybernetic systems. This corresponds quite well to Beer's VSM (Viable System Model).

Situation Coverage

It would be useful to have methods of measuring the "goodness" of architectures. That is, to develop measures and thus be able to establish thresholds. One obvious measurement that can be related to the System Coupling Diagram with respect to Respondent Systems is their degree of situation coverage. That is, how well the Respondent System addresses the problems or opportunities that arise in a Situation System. Further, when system assets are taken from the Enterprise portfolio for incorporation into the Respondent System, some form of measure as to their contributions (as elements) toward assisting the Respondent System in meeting its objective of situation coverage would be useful. To a large extent, this is still an open research question.

There is some promising research in this area being performed at MIT in the SEAri (Systems Engineering Advancement Research Initiative, http://seari.mit. edu). In this work, five aspects of engineering complex systems are considered. In addition to the Structural and Behavioral properties of systems consideration is given to the Contextual, Temporal and Perceptual properties of architectures. Thus, a holistic view of situations is being addressed. Contextual properties are related to circumstances in which the system exists (this relates directly to the earlier discussions of the NSOI, WSOI, Environment and Wider Environment). Temporal properties are related to the dimensions of systems over time (a useful architectural related perspective that the system design must include adaptation for the future). Perceptual that is related to stakeholder preferences, perceptions and cognitive biases (again part of a holistic view). By considering sets of conceptual properties of a system in respect to these five aspects various alternative architectural solutions can be evaluated in respect to "goodness". [Rhodes and Ross, 2010] provide a detailed description of these five aspects and how they can be applied.

Finally, there is an international standard [ISO/IEC 15939] that provides guidance in measurement as well as guidance in establishing measurement indicators from a systems engineering perspective. [Roedler, et. al., 2010]

DECISION MAKING

"It's always wise to raise questions about the most obvious and simple assumptions."
C. West Churchman [Churchman, 1971]

It is important to remember that decisions are made all of the time, at all levels in an enterprise by a line organization, a project organization, or an individual. It is essential to the enterprise that important decisions are made based upon prudent analysis of situations, the identification and evaluation of alternative solutions in respect to the quality, risks and costs involved and that changes resulting from important decisions can be followed up to determine if the desired "Effect" has been achieved.

The ISO/IEC 15288 standard provides the following Project processes to be utilized in relationship to decision making.

Decision Making	select the most beneficial course of project action where alternatives exist
Risk Management	reduce the effects of uncertain events that may result in changes to quality, cost, schedule or technical characteristics
Measurement	collect, analyze, and report data relating to the products developed and processes implemented within the organization, to support effective management of the processes, and to objectively demonstrate the quality of the products

While these processes belong to the Project category, they can be tailored and instantiated wherever they are needed within the enterprise, for example as an integral part of the implementation of the Change Management System.

Change Triggers - Reactive and Proactive Decisions

There are many problem and/or opportunity situations that can trigger the need for a change, for example:

- Changes in the external environment
- Changes in insight within the enterprise
- Changes in the marketplace
- Changes in what technology can offer
- Changes in resource availability (facilities, people, raw materials)
- Changes during the development of system descriptions

– Changes due to consumption (operation and maintenance)
– Changes due to product or service quality

A healthy enterprise responds to change triggers in the short term in a reactive manner, but at the same time continually evaluates its capabilities to achieve its purpose, goals and missions via its infrastructure related system assets in a proactive manner. Such enterprises, when making necessary reactive changes do not loose site of the holistic aspects of their systems and their operation. They continually identify potential sources of problems and opportunities, initiate Respondent Systems in the form of projects, missions or task forces, study thematic systems in order to gain insight, plan for change and establish and monitor Respondent System efforts to carry out change.

An unhealthy enterprise responds to change triggers in a reactive decision making manner. They react to problem situations or jump to seize an opportunity resulting in isolated change decisions that can have a deep and destructive effect upon their infrastructure as well as their value added system products and/or services.

Strengths, Weaknesses, Opportunities and Threats

There are a number of methods and tools that can be applied in supporting decision-making. One very straightforward method is the use of SWOT that was developed at the Stanford Research Institute in the 1960s and 70s and is usually credited to Alfred Humphrey . SWOT is a framework for analysis of strengths and weaknesses, opportunities and threats that can be deployed at any level in an enterprise. It is a very useful tool for the Change Management function. A SWOT analysis assists in focusing upon strengths, minimizing weaknesses, and taking the greatest possible advantage of opportunities available. Thus a SWOT analysis is useful in identifying important factors related to the systems in the Enterprise system portfolio. It provides useful support for the Orient part of the OODA loop as described in Chapter 3. SWOT findings are put into a table of four quadrants as portrayed in Figure 5-4. The strengths and opportunities are considered helpful for achieving an objective; whereas the weaknesses and threats are considered harmful to objective achievement. Further, strengths and weaknesses are related to enterprise internal aspects; whereas opportunities and threats are typically external to the enterprise.

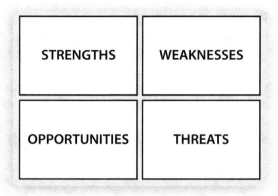

Figure 5-4: SWOT Analysis Format

The following provides typical questions that can be addressed in a SWOT analysis and exemplifies factors that could be included.

Strengths

What advantages do you have?	special marketing competence
What do you do well?	an innovative product or service
What relevant resources do you have access to?	geographical location of your business
What do other people see as your strengths?	high quality processes and procedures
	other aspects of your business that add value to your products and services

Weaknesses

What could you improve?	lack of marketing expertise
What do you do badly?	lack of differentiating products and services (in comparison to competitors)
What should you avoid?	geographical location of your business
	inferior quality products and or services
	damaged reputation

Opportunities

Where are the good opportunities facing you?	developing market such as internet
What are the interesting trends you are aware of?	possibility for joint ventures, mergers or strategic alliances
	moving to new market segments that provide improved profits
	new international market
	market vacated by an inferior competitor

Threats

What obstacles do you face?	new competitor enters your market
What is your competition doing?	price wars with competitors
Are the required specifications for your job, products or services changing?	competitor develops an innovative product and or service
Is changing technology threatening your position?	competitors have better access to distribution channels
Do you have bad debt or cash-flow problems?	taxation of your product or service is introduced
Could any of your weaknesses seriously threaten your business?	

A SWOT analysis can be very subjective and thus should not be relied upon as the only approach to orientation on the enterprise situation. Some guidelines for a successful SWOT analysis are as follows:

- realistic portrayal of enterprise strengths and weaknesses
- distinguish between where the enterprise is today and where it could be in the future
- be specific and avoid fuzzy areas
- analyze the enterprise situation relative to the competition
- keep the analysis short and avoid detailed analysis and complexities

Note: The reader is advised to do a web search on SWOT where additional explanations of SWOT as well as a large number of examples are provided. While this discussion has concentrated upon the enterprise use of SWOT, it is equally valid for individual use in analyzing personal situations.

Decision Trees

Another straightforward method for supporting decision-making is the deployment of decision trees. A decision tree (sometimes referred to as a tree diagram is a support tool that uses a tree model of decisions and their possible consequences, including chance event outcomes, resource costs, and utility. Decision trees are commonly used in decision analysis, to help identify a strategy most likely to reach a goal. Another use of decision trees is as a descriptive means for calculating conditional probabilities. The structure of a decision tree model may differ somewhat, but the portrayal in Figure 5-5 illustrates the principles.

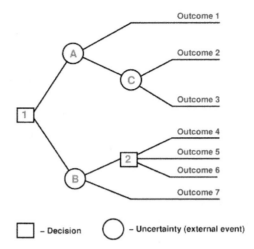

Figure 5-5: Decision Tree Structure
Source: Google: Decision Trees and www.time-management-guide.com

The squares represent decisions you can make. The lines coming out of each square on its right show all the available distinct options that can be selected at that decision analysis point. Circles show various circumstances that have uncertain outcomes; for example, some types of events that may affect you on a given path. The lines that come out of each circle denote possible outcomes of that uncontrollable circumstance. To quantify the trees, annotate above each such line in the decision tree best guesses for probabilities (for example, "70%" or "0.7") of those different outcomes.

Each path that can be followed from left to right, leads to some specific outcome. Describe those end results in terms of your main criteria for judging the results of your decisions. Ideally, assign each end outcome a quantitative measure of the overall total benefit that will receive from that outcome; for example, expressed as a perceived monetary value.

The complete decision making tree provides probabilities of the uncertain events and the benefit measures; that is desirability of each end result. At this stage the tree can be used to provide more specific recommendations on what would be the best choices. In particular, for each choice that can be controlled (at the square decision points), you can calculate the overall desirability of that choice. This is the sum of the benefit measures of all the end outcomes that can be traced back to that choice, via one path or another, weighted by the probabilities of the corresponding paths. This will indicate the preferred choice, that is, the one with the highest overall desirability.

Once again, a web search on Decision Trees will yield several variants of decision tree models and loads of examples.

Consistent Decision Making

One of the most important aspects of decision-making is consistency in respect to established concepts and principles. Concepts and principles for a system to be change managed are reflected in the architecture of the system as described in Chapter 4. In this regard, it is important that the architectural design of the system is based upon a small number of clear concepts and principles. When this is the case, decision making during the entire life cycle tends to stay within boundaries that are explicitly or implicitly implied by the concepts and principles as portrayed in Figure 5-6. Thus cow chasing due to not establishing and communicating driving concepts and principles as described in Chapter 4 can be avoided.

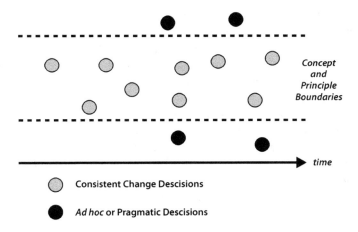

Figure 5-6: Concepts and Principles Bound Decision Making

Decisions that lay outside of the boundaries tend to be ad hoc or pragmatic and diverge from the concepts and principles. The first diverging decision often leads to the second, third and so on. The implementation of such decisions over time leads to the decay of the architectural structure of the system. While many of the system products supplied by the computer industry are definite examples of this phenomenon, there are many other man-made systems that suffer from inconsistent decision-making due to the lack of strong concepts and principles.

The Entropy Phenomenon

Another important system phenomenon that is central to decision making is that of *entropy*. Entropy has its origin in the field of thermodynamics where it is used to describe the measurement of decaying behavior (disorder); namely the effect of the spreading of heat resulting in the loss of energy. However, the notion of decaying behavior of systems is general and can well be applied to both physical as well as other types of man-made systems, as noted in the previous discussions on consistent decision-making [Rifkin, 1980].

Physical systems in nature, like the human body, require inputs in the form of energy in order to sustain operation. If the human does not receive this energy, the body will become anorexic and will eventually die. A similar phenomenon, for example, can be observed for man-made systems like a climate conditioning system that does not receive the energy required to heat and/or cool the environment in which it operates. The decay of such a system from order toward disorder is called *positive entropy*; that is, entropy increases.

Energy of an appropriate form is required to achieve *negative entropy*; decreasing entropy resulting in improved system behavior. For example, the injection of energy to the human body counteracts the entropy effect that leads towards anorexia. The following citation points to the importance of negative entropy for the change management activities of an organization (enterprise):

> *"Most change efforts fail because they are not given enough follow-up, reinforcement and new energy. Many managers want to get everything up and running on autopilot, but this is the antithesis of what actually makes change happen. In systems terms, it takes negative entropy-new energy-to make change occur. In fact, most executives are concerned about getting employee "buy-in," when "stay-in" is even more difficult to get and retain over time."*
> **S.G. Haines** [Haines, 1998].

The lack of negative entropy (new energy) in a system is what leads to obsolescence, rigidity, decay, and ultimately death of a system. Thus, one of the most important requirements of a Change Management System is to hinder positive entropy and to make changes that will lead to negative entropy in the systems over which authority and responsibility are vested. This is not an easy task in respect to the complex interrelationships that exist between multiple systems. A decision to make a change resulting in the injection of required energy in one system may actually have side effects that result in positive entropy for other systems. Remember the discussion of the chain of multiple causal relationships, paradox, and system archetypes containing growth loops from Chapter 2.

The analysis and understanding of complex relationships that lead to positive and negative entropy can be portrayed utilizing the Links, Loops Language as introduced in Chapter 2. Thus, in Figure 5-7, basic relationships of entropy are portrayed in the form of growth archetypes.

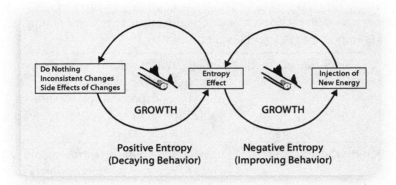

Figure 5-7: Growth Due to Positive and Negative Entropy

In the case of positive entropy, the system decays as a result of no action, inconsistent decisions or as a causal side effect of changes to one or more other systems. This is not to imply that all changes made in related systems will necessarily result in positive entropy. In the case of negative entropy, the change resulting in new energy injected into the system is typically in the form of resources (human, facilities, or financial) but also can occur due to increased individual and/or group commitment and *esprit de core*.

IMPLEMENTATION OF CHANGE MANAGEMENT

How is the change management element of the Change Management System portrayed in Figure 5-3 to be implemented? This is a vital question for the enterprise in which one or more instances of the Change Management System are implemented. Some of the central issues are:

- What processes support change management?
- Who owns change management (system description and instances)?
- How is change management distributed?
- What instruments are required for controlling the life cycle management of systems?

Change Management Processes

What process or processes define the outcomes of and the operational activities of change management as it exercises its controlling role over the management of the life cycle of the systems for which it has authority and responsibility?

The ISO/IEC 15288:2008 standard does not explicitly identify a Change Management Process; however it does provide Organizational Project-Enabling processes that contain important change relevant aspects as follows:

Life Cycle Model Management	define, maintain, and assure availability of policies, life cycle processes, life cycle models, and procedures for use by the organization with respect to the scope of the International Standard
Infrastructure Management	provide the enabling infrastructure and services to projects to support organization and project objectives throughout the life cycle
Project Portfolio Management	initiate and sustain necessary, sufficient and suitable projects in order to meet the strategic objectives of the organization
Human Resource Management	ensure the organization is provided with necessary human resources and to maintain their competencies, consistent with business needs
Quality Management	to assure that products, services and implementations of life cycle processes meet enterprise quality goals and achieve customer satisfaction

In providing the process capabilities required for change management, one solution is to use the ISO/IEC 15288 Organization Project-Enabling processes as a starting point and then tailor them in terms of their purposes, outcomes and activities so that they reflect the organization (enterprise) needs in respect to change management. Another alternative, is to, via tailoring, create a Change Management Process with a purpose such as the following:

Change Management	make decisions related to and provide for the control of changes of any nature that are essential to achieving enterprise purpose, goals and missions

The outcomes of this process as well as the activities should reflect the essential properties of the control element of the Change Management System model that has been presented in this book. Given this process definition, the ISO/IEC 15288:2008 Organizational Project-Enabling processes should also be tailored to work in harmony with the Change Management Process.

The advantage of defining a new process is that it brings focus to the importance of Change Management as a central Organization/Enterprise function. On the other hand, it establishes another process definition that must be managed.

Ownership of Change Management

In Chapter 4, the topic of system ownership was addressed. Since Change Management is a defined human activity system, it has a system description and can exist in multiple instances that provide a vital service throughout an enterprise. The system description of the Change Management System is provided in this book, via the graphic model in various versions as well as corresponding text related to its composition and utilization. As with all systems, the Change Management System must be life cycle managed. The system description of Change Management is a vital infrastructure asset and is owned and life cycle managed by the enterprise. Within the enterprise instances of the Change Management system are owned by the party or parties that utilize (operate) the respective instances. While the name given to the party or parties owning instances of change management may vary, they will be called Change Control Boards (CCBs) henceforth in this book.

Russel Ackoff addressed the ownership question in his view of the circular organization where a participatory structure based upon Change Control Boards is utilized to achieve a form of democratic hierarchy [Ackoff, 1994]. Three principles are put forth by Ackoff to exemplify participation through a structured approach in the form of CCBs.

- The absence of ultimate authority.
- The ability of each member to participate directly or through representation in all decisions that affects him or her directly.
- The ability of members, individually or collectively, to make and implement decisions that affect no one other than the decision-maker(s).

According to Ackoff's organization model, there are six responsibilities of each board, regardless of level.

- To plan for the unit whose board it is.
- To make policy for the unit whose board it is.
- To co-ordinate plans and policies for the immediate lower level.
- To integrate plans and policies with those immediately below it and those at higher levels.
- To improve the quality of work life of the subordinates of the board.
- To enhance and evaluate performance of the manager whose board it is.

While Ackoffs model represents an extreme of democratic decision-making, in reality, the assignment of authority and responsibility for change is vested in the organization management. While dealing with change in a dictatorial manner is often counter-productive, there are varying degrees between this extreme and universal democracy. Some environments, particularly the military, have functioned primarily as a result of absolute authority and responsibility where strict obedience to orders is essential. However, in most situations some form of representation for stakeholders of the systems to be life cycle managed is the most appropriate. The most important CCB goal is to accumulate knowledge from decision making as a part of the learning organization so that the Change Management System and the processes to be implemented within its scope are continually improved.

Distributed Change Management

Change Management Systems are instantiated, managed and operated by CCBs on behalf of an organization (enterprise). The authority and responsibility for change is distributed based upon the systems that are to be life cycle managed by each CCB. A level-wise representation of this is portrayed in Figure 5-8.

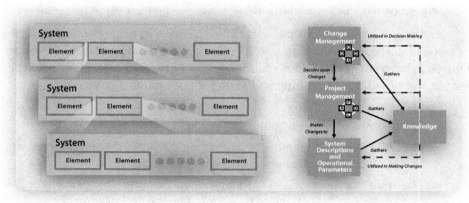

Figure 5-8: Distributed Change Management Systems

At the highest level the CCB operates to coordinate the life cycle management of its System-of-Interest that is composed of system elements that it deems necessary as assets to achieve the purpose, goals, and missions of that level of the enterprise. The CCB at this level delegates authority and responsibility to lower level CCBs to life cycle manage system elements as Systems-of-Interest at their level according to the recursive system decomposition described in Chapter 1. However, as Ackoff proposes, the question of absolute authority may be weakened so that the organization operates in a more democratic manner across organizational levels where even the lowest level can influence change decisions at higher levels. This certainly contributes to achieving a more holistic view of the system structures and the defined interrelationships.

CCBs are a virtual management organization form where membership must represent all the essential aspects of managing the life cycles of the systems over which authority and responsibility is held. At the highest Enterprise level, this can correspond the systems in the aggregate Enterprise system that will be described in Chapter 8. Thus membership in a CCB typically reflects at least the immediate lower level managers as well as experts that can contribute to the decision-making of the CCB. It is possible, and often desirable for long-term projects, to incorporate a CCB within a project wrapper. Such large efforts may be identified as Programs by some organizations.

A complicating factor concerning the distribution of CCBs is the utilization of a supply chain where other enterprises become involved in supplying a complete System-of-Interest or system element (either as system descriptions, instances, or services related to utilization) as portrayed in Figure 4-10. The supplying enterprise may or may not have a decision making body that operates in a CCB like manner. In such situations, the importance of establishing agreements that stipulate how change decisions and changes are to be treated between acquirers and suppliers becomes critical.

Instruments for Life Cycle Management

In order to perform Change Management activities, a number of status related instruments are required. An important instrument is the life cycle model for the System(s)-of-Interest that the CCB life cycle manages. At a minimum some form of matrix describing the contributions of Technical Processes (those that are actually used to accomplish transformations) in respect to life cycle stages is useful in this regard. The generic structure of such a matrix is portrayed in Figure 5-9.

Within each stage, the contributions of each Technical Process to the stage result(s) (if contributing in that stage) are identified. By taking a complete life cycle view for each process, a holistic perspective of changes during the entire life cycle is obtained. Note that it can be useful (even desirable) to provide matrix rows for the contributions of other processes from the Organization Project-Enabling, Agreement, and Project process categories as described in Chapter 3.

From the structure of the life cycle, the CCB can determine the boundaries of projects in respect to both the life cycle stages and the technical processes (the Do part of the PDCA loop) that are relevant for each project. At one extreme a single project may be assigned the responsibility for all life cycle stages and all technical processes. At the other extreme projects may be assigned a single stage, a single process (perhaps even a single activity) within a single stage.

Stages\ Processes	Stage A	Stage B	Stage C	...	Stage X
Requirements Definition					
Requirements Analysis	*Contributions to the Achievement of Stage Outcomes*				
Architectural Design	...	...	...	...	...
Implementation					
Integration					
Verification					
Transition					
Validation					
Operation					
Maintenance					
Disposal					
Stage Results	Outcomes	Outcomes	Outcomes	Outcomes	Outcomes

Figure 5-9: Life Cycle Model Matrix

The outcomes of stages become the results reviewed at Decision Gates as explained in Chapter 3. As noted above, the ISO/IEC 15288 Decision Making, Risk Management and Measurement Processes can be tailored and utilized in implementing this important type of CCB decision-making.

A matrix for each type of man-made system to be life cycle managed by the CCB is required. Examples of life cycles for various types of systems are presented in Chapter 6.

As noted in the discussion of distribution of change management, another important instrument is the agreement. Agreements based upon the usage of the Acquisition and Supply processes can be made with external or internal suppliers of system definitions, production or consumption related products and/or services as illustrated in Figure 4-10. Agreements can be made at the CCB enterprise level, at the project level or within line organizations. The agreements are the glue that holds together the acquiring and supplying elements of a supply chain system.

Management of Processes and Life Cycles

As described earlier, the processes and life cycle models that an organization (enterprise) utilizes are essential infrastructure assets of their systems portfolio. Thus, a major responsibility of change management is making decisions related to the operational capability (readiness) of the organization (enterprise) to meet its commitments in terms of purpose, goals, and missions.

The set of processes that an organization (enterprise) utilizes form a Process System that like all other systems must be life cycle managed. The life cycle management of the Process System may be centralized at the highest-level CCB or distributed to various levels of CCBs.

Likewise, Life Cycle models individually and collectively are System(s)-of-Interest to CCBs and must be life cycle managed. That is, there must be a life cycle model for regulating stages of the set of life cycle models that are appropriate for a CCB at any level.

The change management of processes and life cycle models is one of the primary responsibilities for the CCB. A representation of how the CCB should treat change triggers, utilize tools such as the Decision Trees and SWOT analysis models in making prudent decisions about the enterprise capability to accomplish needed changes is portrayed in Figure 5-10.

In responding to change triggers (reactively or proactively) an assessment of current capability is performed. Based upon the enterprise needs a process profile required to achieve appropriate results is defined. Finally, decisions are made to make the appropriate changes in Processes, Life Cycle models or Project practice. The changes are then made and followed-up by the CCB.

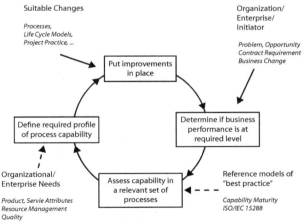

Figure 5-10: Responding to the Need for Changes

Note: The author gratefully acknowledges the contribution of Dr. Jonathan Earthy who provided an earlier version of this figure.

CHANGE MODEL FOR RAPID DECISION MAKING

The need for making rapid change decisions is characteristic of the field of Command and Control. While Command and Control has a long history in military sectors, it also provides a basis for directing operations in various types of civil crises situations (fires, floods, train crashes, terror actions, etc.). The OODA loop of the Change Management Paradigm loop is still vital in such situations. However, it is useful to be more specific about the OODA related functions for the area of Command and Control. Berndt Bremer [Brehmer, 2005] has over several years refined OODA to a model called DOODA (i.e. Dynamic OODA). A slightly modified version of Brehmer's DOODA loop that can be applied in providing the control element of Respondent Systems is portrayed in Figure 5-11.

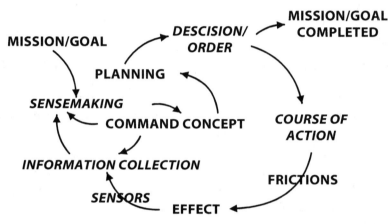

Figure 5-11: Dynamic OODA Loop

All of the elements of the modified DOODA loop expressed in italics are functions to be performed via the application of one or more enterprise system assets. This includes *SENSORS, INFORMATION COLLECTION, SENSEMAKING, PLAN-NING, DECISION/ORDER* and *COURSE OF ACTION*. The other elements of the loop are as follows:

The MISSION/GOAL is the threshold value for the iterative cybernetic DOODA loop and exit from the loop transpires when the decision is made to terminate. However, the MISSION/GOAL COMPLETED state can either be due to success or failure.

The COMMAND CONCEPT provides guidelines for decision-making. This aspect is vital since it sets the boundaries for consistent decision-making according to an established strategy in a manner similar to the importance of architectural consistent decisions as discussed earlier in this chapter.

FRICTIONS are encountered in executing the Course of Action (COA). Frictions reflect the fact that the execution of a COA does not always take place as planned. This can equally apply to Projects carried out in the life cycle management of sustained systems as well as for Task Forces charged with operational responsibility for the COA.

The EFFECT is an observable state that is reached in each DOODA loop cycle. As was discussed in the application of Senge's Links, Loops and Delay language, there may be delays, even long ones, before effects become observable.

Once again, this change management model is a system where the elements and relationships have just been defined. It would be wise for enterprises dealing with command and control like functions to consider "institutionalizing" such a model and place it under life cycle management. In this manner, instances of the model can be "produced" as a service at appropriate times in a respondent system in order to meet in particular, often critical, situation system needs.

Sensemaking

Amongst the functions performed in the DOODA loop is *SENSEMAKING,* which is a relatively new area of study. While the goal of the function is clear in an abstract sense, there are a number of related techniques and methodologies that have bearing upon this topic.

The sensemaking function can be viewed as a paradigm, a tool, a process, or a theory of how people reduce uncertainty or ambiguity or socially negotiate meaning during decision-making events [Ntuen et.al., 2005]. Sensemaking involves the collective application of individual "intuition"—experience-based, sub-consciously processed judgment and imagination—to identify changes in existing patterns or the emergence of new patterns [Weick, 1995]. Through the accurate construction of meaning clarity increases and confusion decreases. A poor sensemaking process often leads to poorly understood objectives, goals, missions, and visions. This in turn can lead to poor framing of plans, and consequently, poor decisions. A peruse of literature on sensemaking can be summarized as follows:

How meanings and understanding of situations, events, objects of discourse, or contextual information are produced and represented in a collective context.

There are a number of methods and tools available for supporting sensemaking which include the following:

Information Fusion – Via the fusion of data and information from sensors and/ or data bases, new data and information is created that can be used to support sensemaking. For example, the fusion can be used to verify the reliability of data and/or information. Alternatively the fusion can result in uncovering new aspects that may not be obvious by examining single data or information sources. As with system properties, by combining data and/or information elements new properties emerge. [Wik, 2003] describes relationships of Multi-sensor Data Fusion and Network Based Defence.

Systems Thinking – The languages and methodologies of Systems Thinking as introduced in Chapter 2 are very relevant for sensemaking. As a discipline for seeing wholes and patterns of change rather than static snapshots it is directly relevant the sensmaking processes. Due to many of the inherent similarities, sensemaking can be considered to be a part of rational systems thinking.

Truth Maintenance – The identification of truth has long been a subject of interest in the field of Artificial Intelligence. Truth Maintenance Systems (TMS), also called Reason Maintenance Systems, are used in Problem Solving. In conjunction with Inference Engines (IE) such as rule-based inference systems, they are used to manage as a Dependency Network the inference engine's beliefs in given sentences [Ingargio, 2005] A TMS is intended to satisfy a number of goals:

A. Provide justifications for conclusions - When a problem solving system gives an answer to a user's query, an explanation of the answer is usually required. An explanation can be constructed by the IE via tracing the justification of the assertion.

B. Recognize inconsistencies - The IE may tell the TMS that some sentences are contradictory. Then, if on the basis of other IE commands and inferences we find that all those sentences are believed true, the TMS reports to the IE that a contradiction has arisen. The IE can eliminate an inconsistency by determining the assumptions used and changing them appropriately, or by presenting the contradictory set of sentences to the users and asking them to choose which sentence(s) to retract.

C. Support default reasoning - In many situations we want, in the absence of firmer knowledge, to reason from default assumptions. If Tweety is a bird, until told otherwise, we will assume that Tweety flies and use as justification the fact that Tweety is a bird and the assumption that birds fly.

D. Remember derivations computed previously – Previously derived conclusions will not need to be derived again.

E. Support dependency driven backtracking - The justification of a sentence, as maintained by the TMS, provides the natural indication of which assumptions need to be changed if we want to invalidate that sentence.

Given these properties of a TMS one can observe that truth maintenance is a knowledge representation method for representing both beliefs and their dependencies.

The name truth maintenance is due to the ability of these systems to restore consistency. Such automated techniques can be essential to sensemaking, in particular in assisting in identifying the presence of inconsistent decisions. However, at this point in time very little research has been done in the coupling of sensmaking and truth maintenance.

DEPLOYING SYSTEM ASSETS

Earlier in this chapter, factors related to the implementation of change management were introduced. The implementation of functions within change management is based upon deploying system assets that can include processes, equipment and personnel in the control element of a Respondent System. The deployment of system assets during command and control operations like DOODA (including courses of action) involves coupling of the assets to functions as portrayed in Figure 5-12.

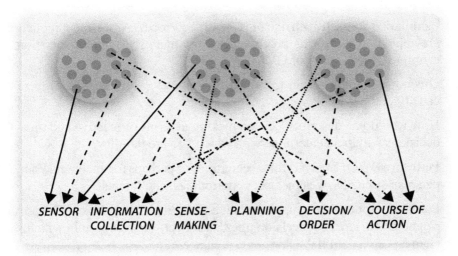

Figure 5-12: Mapping System Assets to DOODA Functions

System assets can be owned and operated by a single enterprise. However, as this figure portrays the system assets of multiple (extended) enterprises can be applied to the functions. In the military context, multiple enterprises may be the result of system assets of various land, sea or air forces being polled to provide a network based warfare capability. In another rendition, it can apply to the assets from various countries involved in a joint operation.

KNOWLEDGE VERIFICATION

1. How are important decisions made in an enterprise with which you are familiar?

2. Describe the advantages and disadvantages of a dictatorial versus a democratic style of decision-making.

3. Apply the Cybernetic System model to some physical and non-physical systems with which you are familiar. Clearly identify the Control, Controlling and Measurement Elements. Also, describe how regulation is achieved, what effects are produced, how measurements are made and which thresholds are utilized.

4. Describe how the OODA and PDCA loops become important elements in viewing the Change Management model as a cybernetic system.

5. Given the system examples from (3) identify appropriate MOEs (Measurements of Effectiveness) and MOPs (Measurements of Performance).

6. Why is customer satisfaction an important effect measurement?

7. Familiarize yourself with one or more of the Process Assessment models that are applied in measuring process capability maturity of an organization; for example CMM, Spice, or CMMI.

8. Given the examples from (3), identify various types of change triggers that can affect the systems.

9. How would you describe decision making in terms of reactive and proactive decisions within an enterprise with which you are familiar?

10. Perform a SWOT analysis for an enterprise or perhaps for analyzing your own personal strengths, weaknesses, opportunities, and threats.

11. Postulate a somewhat complex situation where multiple avenues of responding can be taken. Create a decision tree that reflects the response in which some elements are controllable but some are uncertain.

12. Are system architectural concepts and principles explicitly used in decision making in an enterprise with which you are familiar? If so, what are they?

13. Give some examples of positive entropy (decaying behavior) and negative entropy (improving behavior) in some physical and non-physical systems; perhaps the ones identified in (3).

14. Examine the Organization Project-Enabling processes from ISO/IEC 15288 and determine how they could be tailored to incorporate the needs of a Change Management System as described in this book.

15. Describe problems that may be encountered when changes are to be made by the suppliers in a supply chain.

16. How are projects established and bounded in an enterprise with which you are familiar? What types of mechanisms are used to specify their work?

17. Describe how the DOODA loop can be used for decision-making in crises situations within an enterprise and amongst enterprises.

Chapter 6
Life Cycle Management
of Systems

Bringing the pieces together!!!

In the previous chapters many essential aspects of man-made systems have been described. In this chapter, the knowledge that has been gained is put to use in further describing various aspects of the life cycle management of systems.

MANAGEMENT AND LEADERSHIP OF SYSTEMS

Harvard professor John Kotter identifies differences between management and leadership:

> "Leadership and management are two distinctive and complementary systems of action. Each has its own function and characteristic activities. Both are necessary for success in a complex and volatile business environment"
> **John Kotter** [Kotter, 1990].

Some tone setting differences identified by Kotter are:

Setting a Direction vs. Planning and Budgeting

Aligning People vs. Organizing and Staffing

Motivating People vs. Controlling and Problem Solving

Thus, management is primarily concerned with the near-term day-to-day operation of an enterprise. Within this context, there are obviously many types of changes that must be considered, however, they tend to be changes in the operational parameters associated with the enterprise infrastructure systems that are in operation. Leadership on the other hand, is involved most often with making longer-term strategic structural changes in the systems that will lead to new behaviors (effect) when they are put into operation. Such leadership is based upon well thought out visions of the future.

The change management system presented in this book includes the treatment of changes to system descriptions and to operational parameters. At a first look, this delineation points to the fundamental difference between managing systems (via changes in operational parameters) versus leadership of systems (changes in fundamental description, i.e. structure). However, the delineation is not that simple since the ability to change system structure may also exist within the jurisdiction of near-term activities. Likewise, making operational changes may indeed be of such strategic importance that true leadership may be required in order to assure that near-term operative changes do not have a negative impact upon long-term strategic goals.

A Change Control Board (CCB) is a fundamental decision-making organ of change for system descriptions and operational parameters. While the board functions as a collective group, the CCB must empower; that is transfer authority and responsibility to individuals or small teams for continued "local" near-term or long-term decision-making activities. In the empowerment, clear bounds must be established so that the purpose, goals, and missions of the enterprise are not compromised by improper, inconsistent decision-making as described in Chapter 5. Once again, we return to the System Coupling Diagram that is applicable both for management and leadership activities as portrayed in Figure 6-1.

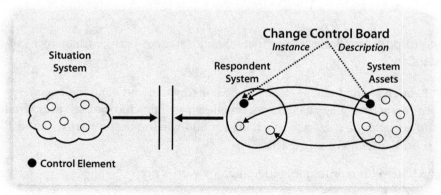

Figure 6-1: Change Control Board Asset (Description and Instance)

Here we observe that a Change Control Board is described as a System Asset that is placed in the Enterprise portfolio. The architecture of the CCB, as with all other systems reflects the enterprise policies, needed capabilities, requirements as well as for this type of human activity system, the structure for handling the flow of data and information and decision-making. For varying situations, the CCB definition may be tailored and then instantiated thus becoming the control element of a Respondent System.

Two important factors that determine the degree of management versus leadership of systems are the longevity of the systems and the type of value-added products and services that the organization (enterprise) produces.

System Lifetimes

Some situations that are handled via Respondent Systems have a very short lifetime (for example, a week, a day, an hour, or even a few seconds) and thus require a different form of management and leadership than those that live for months, years or decades.

System products and services that are produced in response to the needs of some type of "marketplace" that have long longevity are often upgraded when new technology and/or new knowledge becomes available. Good examples of such long lifetime products are the DC-3 commercial aircraft and the global telecommunication system. Within military circles the investment in system technology and the need to reduce costs often results in the reuse of a system product structures in order to provide new services over many years. Therefore, product and service (system) sustainability certainly is a factor to be considered for long-term systems and requires both management and leadership attention. Even for defined abstract and defined human activity systems with long lifetimes, the fundamental structures used for organizations, enterprises, projects, documentation, agreements, and so on, can remain intact over a long period of time with only small changes being made based upon new problem and opportunity situations as well as new knowledge.

In contrast to these long-lived systems, there are many examples of systems that are assembled (integrated) in the short-run as Respondent Systems to meet a specific goal or mission. For example, consider a fire brigade put together with the mission to fight a fire. This mission certainly has all of the characteristics of a defined system where system elements are quickly identified and configured to work in an effective manner in the process of putting out (or at least controlling) a fire. In the military, and increasingly in civil contexts, the authority making decisions related to such short-term system efforts is based upon Command and Control (C&C) such as the DOODA loop as described in Chapter 5. In fact, C&C functions in a manner similar to a Change Control Board albeit under significant time pressure. In principle, a C&C makes operative decisions and most often changes parameters to meet the situation involved. Examples of this include, number of fire companies, type and quantity of fire fighting equipment and available personal to be deployed.

There is a strong coupling between the change management of Sustained Assets and Respondent Systems that is often not taken into account. This relationship often falls between the chairs of those people having authority and responsibility for system assets and those having authority and responsibility for their utilization in Respondent Systems. It is quite clear that the capability of an enterprise to operate effectively is dependent upon the state of its system assets. Their readiness for instantiation and operation is vital. Thus, it is essential to assure that the sustained system assets to be utilized in Respondent Systems have been prudently life cycle managed and that their operational performance when required can be guaranteed.

Type of Value Added Products/Services

The common purpose of all organizations, public or private, for profit or non-profit is to produce, via enterprise(s), some form of value added as portrayed in the supply chain relationships of Figure 4-10. The nature of the added value is either a product or a service, or both. In either case the product or service is produced via provisioning of the elements of the system and the integration of the elements according to the system description into a product or service. Various management and leadership approaches can be required based upon the type and complexity of value-added product and/or service that is produced and the type of enterprise involved in the production. Some examples are as follows:

- A *manufacturing enterprise*, for example one producing nuts, bolts and lock washer products sells their products as value added elements to be used by other enterprises who integrate these products into their more encompassing value added system; for example, an aircraft or an automobile.

- A *wholesaling or retailing enterprise* provides products that it offers to their customers. Their customers (individuals or enterprises) acquire the products and use them as elements in their systems.

- A *commercial service enterprise* such as a bank sells a variety of "products" as services to their customers; for example, current accounts, savings accounts, loans, investment management, and so on. These services add value and are incorporated in customer systems of individuals or enterprises.

- A *governmental service enterprise* provides citizens with services that vary widely from health care, highways and roads, pensions, police, defense, and so on. Where appropriate, these services become infrastructure elements that are utilized in larger encompassing systems that are of interest to individuals and/or enterprises.

– A *consulting enterprise* via its services adds value in the form of knowledge and know-how for its customers. For such an enterprise, the set of services "produced" may remain stable for some customers but can also change rapidly as agreements with new customers are established and as customer agreements are terminated.

– An *IT service enterprise* provides data processing and information access capability by operating computers, communication equipment and software systems.

– A *software development enterprise* provides software products that meet stakeholder requirements (needs) thus providing services to product users. Both the developed software and the operation service become part of the set of infrastructure systems of the user enterprise.

Within these examples, there are systems that remain stable over reasonably long periods of time, but there are many that change rapidly. Thus, as stated above the approach to management and leadership must take into account of the type of systems involved as well as their longevity. In turn, the management and leadership directions impact the type and number of life cycle models that are deployed as well as the processes that are made available for utilization within a life cycle.

Missions, Projects and Programs

Earlier it was pointed out that missions and projects are indeed Respondent Systems that are established to handle a Situation System. Both organizational forms have objective (goals) to be accomplished most often within a discrete period of time. However, the granularity of time is more typically in the range of hours, days, weeks, months, or perhaps a small number of years. The system elements of a project include the resources (human, financial, and facilities) provided, the processes to be applied in executing the project, the means of measuring the results of the projects, and so on. Projects provide a prime example of empowerment where authority and responsibility is often transferred to a project manager.

Another form of Respondent System configuration is a program organized to achieve long-term purpose, goal, and mission situations. Programs are operated under the jurisdiction of a program manager, although due to the long-term nature, programs require vision and thus authority and responsibility may be passed on to program leaders. In fact, most programs enlist the services of some form of board of experts that operate as a CCB for the program.

System-of-Systems and their Ownership

In the recursive decomposition of a system as described in Chapter 1, it was observed that system elements of a higher level are systems themselves. Thus the notion of System-of-Systems is natural in the context of decomposition. Furthermore, it has been assumed that each system within the system hierarchy has an owner as described in Chapter 4. However there are an increasing number of very complex systems for which it is often difficult to identify a specific owner [Schumann, 1994]. These have been identified as being a System-of-Systems called a SoS. This was exemplified in Chapter 1 (Figure 1-12) where various actors became involved in responding to events (individually) or crises (collectively). The United States Department of Defense has adopted the following definition:

> *"System of systems engineering deals with planning, analyzing, organizing and integrating the capabilities of a mix of existing and new systems into a system of systems capability greater than the sum of the capabilities of the constituent parts."* [DoD, 2004]

[Boardman, et.al., 2005] provide a clarification of the difference between a generic "System-of-Interest" and a "System-of-Systems," via the following definition:

1. A SoS consists of an assemblage of systems, each system is capable of – and usually practicing – a separate, independent existence;
2. The individual component systems are self-sustaining and purposeful without the SoS;
3. The systems comprising the SoS can be connected and reconnected to produce different, and often unanticipated, effects; and
4. The SoS exhibits the property of "emergence," which means that there are properties characteristic of the SoS that are irreducible and not evident in the component systems.

For example, the Internet can be classified as a SoS since it meets all four points of the definition. Specifically, the individual computers and routers that comprise the World Wide Web can operate quite successfully in a "stand alone" fashion, yet when connected create various effects not evident in the component systems. Another example of a SoS is the North American air traffic control system, whose radar systems, airport systems, and airplane systems can operate effectively either as members of the SoS or independent of one another. In contrast, the gas turbine engine on a B-747 is not a SoS, since it does not does not provide the thrust function independent of the airplane system of which it is an essential part. Similarly, the human heart, while a remarkable system in its own right, cannot effectively function independent of the human body system.

While "owners" of the element systems can be readily identified, who has ownership of the overall SoS, particularly when the SoS transcends organizational boundaries? Such systems are the result of the operation of an extended enterprise [Fairbairn and Farncombe, 2001]. An extended enterprise requires that the organization's entire value chain is addressed, some elements of which transcend the boundaries of what is normally thought of as internal to that organization. In other words all constituent systems including workers, managers, executives, business partners, suppliers, and customers must be considered.

When the ownership of a SoS is fuzzy, it naturally causes problems for both the short-term management and long-term leadership of system life cycle management. While it may not be easy to achieve, some form of CCB (Change Control Board) should be established. However, their jurisdiction and potential to control the owners of the constituent systems may be limited.

LIFE CYCLE MODELS REVISITED

The T-Model

The role of life cycle models was presented and illustrated in Chapters 3 and 4. In Chapter 4, the perspective of viewing stage work products provided by process execution as versions of a System-of- Interest was introduced. Also, in Chapter 4, fundamental changes that take place during the life cycle of any type of defined man-made system were identified to include Definition, Production and Utilization. In Chapter 5, the use of life cycle models as a control instrument for stage result evaluation (decision gates) was described. Building further upon these fundamental properties as well as to facilitate thinking and acting in terms of systems, it is useful to consider the structure of a generic life cycle stage model for any type of System-of-Interest as portrayed in Figure 6-2.

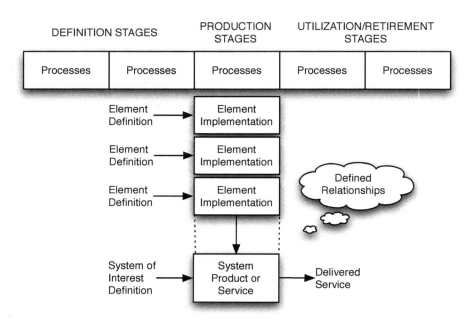

Figure 6-2: Generic (T) Stage Structure of System Life Cycle Models

The (T) model indicates that one or more Definition stages precede a Production stage(s) where the implementation (acquisition, provisioning or development) of two or more system elements has been accomplished. Also in these stages the system elements are integrated according to defined relationships into the System-of-Interest. The Implementation and Integration processes are followed in providing the primary stage results; namely an assembled system product or service instances. Following the Production stage a Utilization Stage is entered. Further relevant stages can include Support and Retirement. Note that this model also displays the important distinction between definition versus implementation and integration.

As noted, this structure is generic for any type of man-made System-of-Interest to be life cycle managed according to ISO/IEC 15288. The Production stage thus becomes the focal point (T) at which system elements are implemented and integrated into system product or service instances. For defined physical systems this is the point at which product instances are manufactured and assembled (singularly or mass-produced). For non-physical systems, the implementation and integration processes are used in service preparation (establishment) prior to being instantiated to provide a service. For software systems, this is the point at which "builds" that combine software elements into versions, releases, or some other form of managed software product are produced.

As described in Chapter 1, using recursive decomposition the implementation of each system element can involve the invocation of the standard again at the next lowest level thus treating the system element as a System-of-Interest in its own right. Thus a new life cycle structure is utilized for the lower level System(s)-of-Interest. When decomposition terminates according to the practical need and risk/benefit related stopping rule, system elements are then implemented (acquired, provisioned, or developed) according to the type of element involved.

As mentioned, for software systems, entry into the Production stages is the point at which "builds" that combine software elements (code modules) into versions, releases, or some other form of managed software product are created. The major difference between systems in general and software systems is the slight variant of the generic model as presented in Figure 6-3.

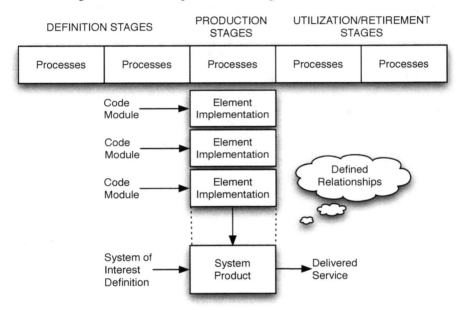

Figure 6-3: T-Model for Software Systems

Stage Execution Order

The most straightforward execution order for life cycle stages is sequential. That is the stages are simply executed in sequence starting with definition stages includes proceeding to production (implementation) related stages and then to utilization. The sequential execution of stages is often referred to as the waterfall model and was first described by [Royce, 1970] in presenting concepts of managing large software systems and later modified by various authors. In fact, for all but the simplest of systems, the waterfall model is not viable. Royce pointed to this lack of viability and presented alternative concepts.

Various types of complex systems require that the stages of the life cycle model be revisited as insight (knowledge) is gained as well as to handle changing stakeholder requirements. Thus, within the context of the (T) stage model, various orderings of stage execution reflecting forms of non-sequential stage ordering can be conveniently described as portrayed in Figure 6-4.

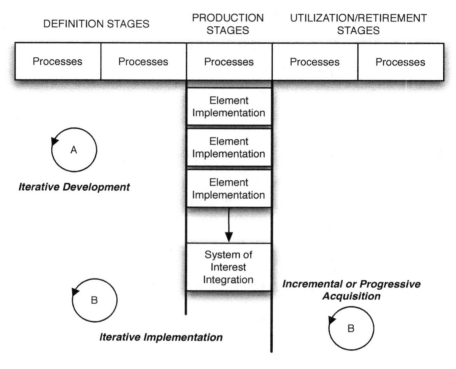

Figure 6-4: Iteration through Life Cycle Stages

Each of the patterns of stage execution involves iteration where previous stages are revisited. The heavy lines denote the demarcation of the iteration end points. Three iterative forms for which several variants can be extracted are:

A. Iterative Development is quite frequently deployed in order to assess stakeholder requirements, analyze the requirements and develop a viable architectural design. Thus, it is typical that a Concept stage (and perhaps a Feasibility Stage as illustrated in Chapter 3) is revisited from a Development stage. For systems where products are based upon physical structures (electronics, mechanics, chemicals, and so on) it is important to get it right before going to production. The need to iterate after production has begun can involve significant costs and schedule delays. Thus, the early stages are used to build confidence (verify and validate) that the solution works properly and will meet the needs of the stakeholders. Naturally, such an approach could be used for software and defined human activity systems as well, however due to their soft nature, it can be useful to go further by experimenting and evaluating various configurations of the system as noted in B.

B. <u>Iterative Development and Implementation</u> involves "producing" (defining, implementing and integrating) various versions of the system, evaluating how well they meet stakeholder requirements (perhaps in the context of changing requirements) and then revisiting Concept (perhaps Feasibility) and Development stages. Such iterations resulting in "builds" are typical within software system development where the cost of "production" is not a significant factor as in the case of defined physical systems. A variant of this approach is the Spiral Model where successive iterations fill in more detail. See [Boehm, 1998]. The use of this approach requires careful attention to issues related to baseline and configuration management as described in Chapter 4. In this approach, significant verification (testing) should be performed on software systems in order to build confidence that the system delivered as a product and/or service to customers will meet their requirements.

C. <u>Incremental or Progressive Acquisition</u> involves releasing systems in the form of products and/or services to the consumers. This approach is appropriate for systems whose structure and the capabilities (functions) to be provided will change in a controlled manner after deployment. The use of this approach can be due to not knowing all of the requirements at the beginning leading to Progressive Acquisition/Deployment or due to a decision to handle the complexity of the system and its utilization in increments; namely Incremental Acquisition. These approaches are vital for complex systems in which software is a significant system element. Each increment involves revisiting the Definition related and Production stages. The utilization of these approaches must be based upon well-defined agreement relationships between the supplying and acquiring enterprises. In fact, the iteration associated with each resulting product and/or service instance may well be viewed as a joint project with actor roles being provided by both enterprises.

In all of the approaches it is wise to use modeling and simulation techniques and related tools to assist in understanding the effect of changes made in the complex systems being life cycle managed. These techniques are typically deployed in the earlier stages however they can equally well be used in gaining insight into the potential problems and opportunities associated with the latter stages of utilization and maintenance (for example, in gaining insight into the required logistics and help-desk aspects).

Allocating and Meeting Requirements

Regardless of the form of stage execution model deployed, stakeholder requirements for the system, including changed requirements in each iteration must be allocated into appropriate activities of the processes used in projects for various stages as well as to the properties of the elements of the system and their defined relationships. The allocation of requirements is performed in the Definition Stages as portrayed in Figure 6-5.

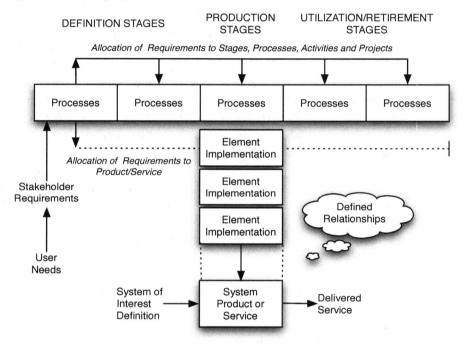

Figure 6-5: Allocation of Requirements

The requirements that are allocated must also be verified to ensure that the work done in the various stages has met the requirements. Furthermore, when a system product or service is transitioned into utilization, it is important to validate that the product or service actually satisfies the User Needs and Stakeholder Requirements. Thus, it can be appropriate to include Verification processes in each stage were specific requirements have been allocated. Furthermore, it is essential to apply a Validation process in the Utilization stage.

Verification	confirm that the specified design requirements are fulfilled by the system
Validation	provide objective evidence that the services provided by a system when in use comply with stakeholders' requirements

Implementation of System Elements

Guidance concerning particular implementation technologies to be applied in an Implementation process is not provided in ISO/IEC 15288. Such implementation technology guidance can be provided by other standards for the specific type of system element to be implemented. For example, standards related to electronics, mechanics, chemicals, software, human factors, product safety, or information security. In several areas, publications presenting "best practice" experience in an area can be used instead of or in addition to standards that provide useful implementation guidance.

For software system elements the ISO/IEC 12207 can be utilized to implement and life cycle manage the software. Alternatively, or in addition, a commercial product such as RUP (Rational Unified Process) could also be applied for guiding the implementation of a software element. Best practice guidance for software service management can be obtained from ITIL publications provided by the UK Office of Government Commerce (GC) and the British Standards Institute (BSI).

It can be advisable to reutilize ISO/IEC 15288 (perhaps complemented with other standards and best practice) for the software element which when considered at the next level of recursion is indeed a System-of-Interest. In this case, the life cycle models and the tailored processes respond to the appropriate needs of software type systems. This approach has the advantage of providing a uniform view of the life cycles of systems thus promoting a common learning organization culture.

PORTRAYING ADDITIONAL ASPECTS OF LIFE CYCLE MODELS

The T-Model has assisted in portraying the essential properties in both the system description, system product and system service dimension as well as in the life cycle, stage, process and project dimensions. Thus, it describes both what and the how. There are other well-known models that have been developed and utilized in describing life cycle related aspects of systems that provide additional insight into life cycle management.

The Vee Model

The Vee Model was developed simultaneously in Germany and the USA. The German V-Model was originally developed in order to describe the development processes of software; whereas the USA version was originally developed for satellite systems involving hardware, software, and human interaction (see Wikipedia, Vee Model) as well as [Forsberg, et. al., 2005].

The T-Model does not portray the relationship amongst processes and their execution in developing the abstract descriptions of systems of interest and their elements. Nor does it portray the concrete development of elements, their integration, verification and validation in providing the value added service expected from the system of interest. To capture these aspects an ISO/IEC 15288 inspired variant of the Vee-Model is provided in Figure 6-6.

The left hand side of the Vee starts with a User Need (not typically expressed in Vee Models). From the need Stakeholder Requirements and Requirements Analysis processes are executed in order to determine the capabilities to be provided as well as the functional and non-functional requirements. The Architectural Design process is executed in order to explore architectural solutions and select one as the desired solution. As part of this process, the Elements of the system-of-interest are defined as indicated at the bottom of the left part of the Vee. As noted these processes can be and often are iterated in order to pin down a solution to the system-of-interest that satisfies the need as well as the stakeholder requirements.

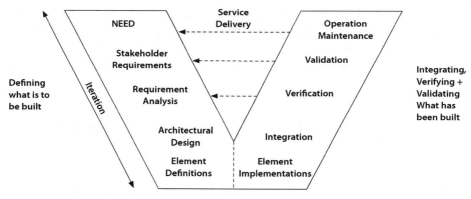

Figure 6-6: A Vee Model Representation of Process Execution

Thus, the left side of the Vee defines what is to be built. On the right side going up the Vee the Elements of the system-of-interest are built according to the Implementation process. This can result in the acquisition of elements or when the element itself is to be developed and produced as a lower level System-of-Interest according the notion of recursion. The System Product or Service to be provided is the result of the Integration, Verification and Validation processes. Finally, the Product or Service is put into operation and maintained in providing the service delivery required in order to meet the User Need.

As noted, this version of the Vee model was constructed to illustrate ISO/IEC 15288 process execution. It would have also been possible to specify the left and right side of the Vee Model by identifying the various system versions (capabilities, requirements, functions or objects, product and service) as indicated in life cycle model introduced in Chapter 4. In the book by [Forsberg, et. al., 2005] several variants of the Vee Model are identified and constructively utilized in emphasizing various aspects of system definition and construction. For example, how baselines are treated in respect to the left and right side of the Vee in respect to definition, verification and validation.

The Spiral Model

While the T-Model portrays the iterative nature of the application of life cycle stages and processes, it does not capture the true effect of the partial transformations that take place during the iterations. For this portrayal, the Spiral-Model [Boehm, 1988] is superior. This model was developed primarily to describe the iterative development of software systems. The version of the Spiral-Model presented here is inspired by the original model, but focuses upon life cycle transformations for systems in general. In this respect it uses the successive life cycle transformations given in Figure 4-1 and is portrayed in Figure 6-7.

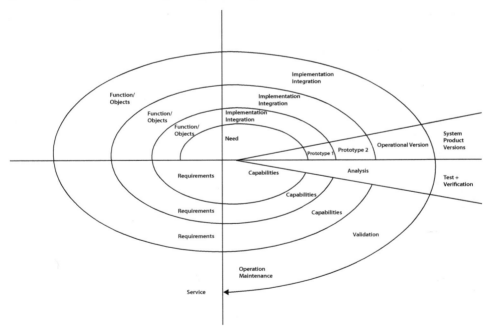

Figure 6-7: Spiral Model of System Transformations

In this rendition of the spiral model the system work products developed during the life cycle are identified. For systems prototype versions of the system product are defined, implemented and integrated in each spiral. After testing and verifying the system product against requirements, the results are analyzed and used as input to the next iteration. The work to be performed the iterations can involve changes to the system of capabilities, system of requirements, and/or the system of functions or objects; that is the definition of the System-of-Interest. One or more of the system elements are redefined and thus must be re-implemented and integrated in order to produce the next prototype version of the system product. When the test and verification is performed and the results analyzed indicating that the system product can provide the system of services, it is validated and transitioned into operation and maintenance in order to provide the service that meets the user need.

LIFE CYCLE ROLES AND RESPONSIBILITIES

In order to avoid the "over the wall" partitioned thinking evident in many enterprises, it is vital to insure that all actor categories are represented and participate in life cycle changes made to a System-of-Interest. The ISO/IEC 15288 via the life cycle approach promotes the unified interdisciplinary thinking required to achieve the integration of various perspectives during the entire life cycle as portrayed in Figure 6-8. In this figure, the life cycle stages illustrated in [ISO/IEC 24748-1, 2009] are utilized. There could be more or fewer stages involved for any particular System-of-Interest life cycle.

Organizational Roles Contribute to	LIFE CYCLE STAGES					
	CONCEPTION	DEVELOPMENT	PRODUCTION	UTILIZATION	SUPPORT	RETIREMENT
CONCEIVERS	Needs, Concepts, Feasibility					
DEVELOPERS		Engineering, Solutions, Practicability				
PRODUCERS			Fabrication, Assembly Verification			
USERS				Operation, Usage, Validation		
SUPPORTERS					Installation, Maintenance, Logistics	
RETIRERS						Engineering

Consistency Practicability Viability

Cohesion Integrity Feasibility

ACTORS

Integrated Project Team

Through Life Management

6-8: Actor Participation in Life Cycle Activities

Actors involved in various organizational functions (roles) are represented according the categories of their organizational function (role) on the left. The contribution of various actors is denoted in moving across the columns of the matrix. In the column associated with their primary function, they perform (Do) their most central process related activities in order to progress in achieving process, project, and stage results. However, as noted in earlier stages, all actors are involved in planning and contribute to stage results by assuring cohesion, integrity and feasibility of the results (from their perspective). For example in the first stage, other actors in their respective roles would, via appropriate planning, answer questions such as:

Is it Developable? Is it Producible? Is it Usable? Is it Maintainable? Can it be Disposed?

Such vital questions continue to be asked as progress is made from Stage to Stage. When the stage results have been achieved and the life cycle progresses to later stages, the actors involved in earlier stages continue to play a vital role in assuring that the transformations made at each stage do not compromise the solution with respect to consistency, practicability, and viability. Further, as the system is utilized and maintained (perhaps even disposed of) they may ask questions such as:

Where the Concepts followed? Where the Concepts Correct?

Were important requirements missed?

Was an appropriate Architecture utilized? Are there better solutions?

Did production result in reliable, cost effective products and/or services?

Did the operation of the system-of-interest provide expected "value-added"?

Was the product or service maintainable?

Could the product or service be conveniently retired?

The answers to questions such as these provide the basis for gathering Knowledge as a vital part of the Change Management System model presented in this book.

Given such an interdisciplinary approach to life cycle management, it is logical to organize work into Integrated Project Teams (IPTs) where representation of the various actor roles is provided. An IPT may be assembled and work together as a single project during the entire life cycle or major parts thereof. However, IPTs may be assembled to deal with individual stages where such a partitioning of life cycle work is deemed to be appropriate.

The reader will note the similarity between the matrix of Figure 6-8 and the matrix of Figure 5-9 where a general approach to organizing life cycle models, in particular for Technical Processes was presented. In Figure 5-9, corresponding processes involved were listed on the left instead of actor roles. The matrix view conveys the partitioning of projects over stages and the selection of technical processes to be applied by the actors in accomplishing stage results and project plans. Thus, in formulating and supporting projects the CCB should take into account the interdisciplinary advantages of using the Integrated Project Team approach as portrayed in Figure 6-8.

INTEGRATING LIFE CYCLE MODELS AND PROCESSES

As the final step in putting the pieces together, factors involved in managing life cycle models and processes are considered. This is followed by a description of how elements for life cycle model schema and process schema are selected, and then complemented with elements providing practical details in order to form a described life cycle model instance for use in a project or line organization.

Managing Life Cycle Models

In practice, an enterprise should establish (define) a small number of fundamental life cycle structural models that serve as schemas for facilitating "running the business" whatever the nature of their value added products and/or services. The life cycle models are themselves system assets composed of stages as system elements. Thus, the models have life cycles and viewing them in this manner provides, once again, for the type of structural consistency that is important to achieve. The most appropriate owner of the system of life cycle models is the CCB in its role as representing the enterprise interests.

Consider once again the generic (T) model of life cycle stages, system elements, their defined relationships and integration into a system as portrayed in Figure 6-2. It should be now be obvious that when a life cycle model is the system that its elements are Stages and that there are defined relationships between stages. Thus, the Stage model utilized is the life cycle model for managing life cycle models.

Process Management

The processes of the ISO/IEC 15288 standard described in Chapter 3 provide, similar to life cycle models, schema descriptions to be tailored in order to accommodate organizational, enterprise, project, and agreement needs. Within an organization (enterprise) process management is accomplished via the following process:

Life Cycle Model Management	define, maintain, and assure availability of policies, life cycle processes, life cycle models, and procedures for use by the organization with respect to the scope of the International Standard

As noted previously, the Change Control Board (CCB) element of the Change Management system operates on behalf of the enterprise. Thus, the CCB should establish and also life cycle manage, according to its needs and its policies, an environment of desired process schema descriptions that can be supported by an infrastructure of methods, procedures, techniques, tools and trained personnel.

Within a project or line organization directed to make changes to system descriptions or operational parameters, the elements of the established environment are selected and further tailored as required as a part of planning (Plan). The selected activities are then executed (Do) in order to provide changes to system descriptions or in operational aspects, the results are evaluated (Check) the results and efforts are redirected if required (Act).

Tailoring the processes of the standard can also be required in facilitate agreements between an acquirer and a supplier. In this case, in reaching agreement the parties select, structure, employ and perform the elements of an established process to provide products and services. The agreed upon processes may also be utilized to assess conformance of the acquirer's and the supplier's performances with the agreement.

All three forms of tailoring lead to managed processes that are described in terms of the contributions they make to a life cycle model to be deployed by the CCB, by the project or in relationship to an acquirer-supplier agreement. As described in Chapter 3, as a result of the successful implementation of the Tailoring Process:

- A life cycle model is defined in terms of stages and the contributions they make.
- Individual life cycle stages that influence the fulfillment of an agreement to supply a system product or service are described.
- Modified or new system life cycle processes are defined.

As with the set of life cycle models, the processes that evolve from the result of tailoring form a system of processes that are to be viewed as a System-of-Interest and thus life cycle managed by the CCB (the owner of the system). Tailoring can be viewed as a part of the early definition stages of the life cycle resulting in an instance of the process system together with an instance of a life cycle model. The processes after implementation and integration into the life cycle model are put into productive use, maintained, and perhaps eventually retired.

Life Cycle Model Instances

The result of the Production stage leading to the establishment of a life cycle model instance is portrayed in Figure 6-9. Once again, this corresponds directly to the general life cycle model structure for any type of system as described in the (T) model of Figure 6-2.

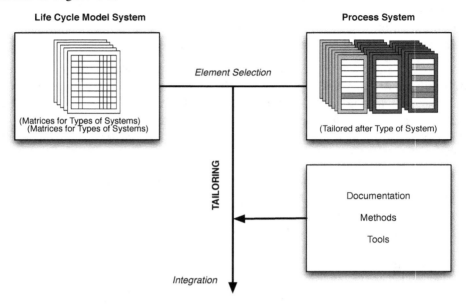

LIFE CYCLE MODEL AND PROCESS INSTANCE TO BE APPLIED

Figure 6-9: Tailoring as the Production of a System Instance

The life cycle model and process instance that are passed on to projects and line organizations become guidelines and control instruments. Thus, ownership is passed from the owner of the definitions (the CCB) to the project or line organization leadership. The transfer occurs in correspondence to empowerment of authority and responsibility.

Provision of Specific Detail, Methods and Tools

The ISO/IEC 15288 standard does not provide specific guidance for documentation, methods, and supporting tools for processes. These detailed elements are to be provided by the users that apply the standard as noted in Figure 6-9. In particular, documentation is always a vital issue. Various types of forms, checklists, etc. are to be applied where deemed necessary. However, documentation should follow the principle of "value-added"; that is, documents should not be required that do not add value and that will not be utilized by the actors in their life cycle roles.

In summarizing, it is through the management of life cycle models and the provision of appropriate sets of processes that an enterprise can establish a uniform means of controlling, executing and measuring the effect of changes to enterprise systems as described in Chapter 5. Furthermore via Knowledge feedback provided from the application of the life cycle models and related processes the effectiveness of the models and processes can be measured and utilized in their change management. This is the necessary path to continuous improvement leading to management and long-term leadership in a learning organization.

PRODUCT LIFE CYCLES

The provisioning of products and services is of course of strategic importance to an enterprise and therefore reflects both leadership and management aspects. The authority and responsibility for the products and/or services is often delegated to the marketing and/or planning functions within the enterprise. They utilize product life cycle management in planning and executing business processes, for example those relating to marketing, sales and support of the products and/or services. The typical life cycle for products (and for many services) is presented in Figure 6-10.

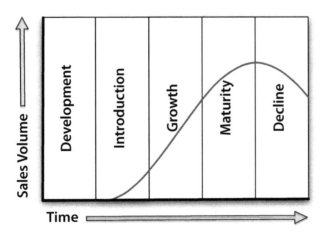

Figure 6-10: Typical Product Life Cycle Management Stages.
Source: Wikipedia: Product Life Cycle Management [www.blurtit.com]

Product life cycles are related to system life cycles where the Development stage of the product life cycle results in the invocation of a system life cycle where the need and requirements for the product are provided to an enterprise development function. The development function defines the system needed to provide product instances. When the product is defined and developed and production begins, the Introduction stage is entered in order to launch the product. The sales volume may fluctuate as indicated in the figure during the Growth, Maturity and Decline stages.

Some products (and services as well) are simply terminated when the market weakens. However certain products and may be renewed in new versions and thus the product cycle is revisited with the next generation of the product or service. Once again, this causes a re-iteration of the system life cycle. For products requiring significant operational service there is of course an after market that often results in a significant source of income for the enterprise, for example in aircraft or automotive products. There may also be contractual and/or legal requirements placed upon the enterprise to provide continued support for the product.

A significant aspect of Product Life Cycle Management is the provisioning of supporting systems that are vital in sustaining operation of the product. While the supplied product or service may be seen as the NSOI (Narrow System of Interest) for an acquirer, the acquirer also must incorporate the supporting systems into a WSOI (Wider System of Interest). These supporting systems should be seen as system assets that when needed are activated in responding to some situation that has arisen in respect to operation of the NSOI. The collective name for the set of supporting systems is the Integrated Logistics Support System (ILS). Some typical types of ILS supporting systems are indicated in Figure 6-11.

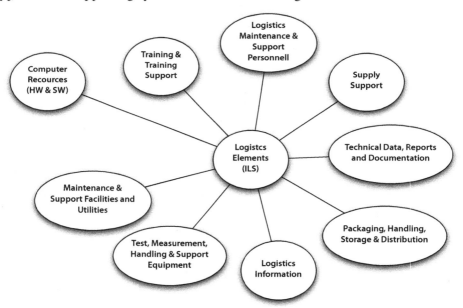

Figure 6-11: Typical ILS Supporting Systems
Source: [Blanchard, 2004]

Consistent with our earlier discussion of System of Interest, the ILS portrayed in the figure identifies several typical elements of this ILS system each being systems requiring life cycle management. The elements are system assets for a supplying enterprise that are instantiated and put into operation in responding to logistics related situations.

As has been emphasized several times earlier, it is vital to have a holistic view when defining, producing and operating system products and services. Thus, in developing the System of Interest, the logistics aspects must be taken into account early in the life cycle. They indeed place vital requirements on the system. In Figure 6-12 the relationship between system design and development and the logistics requirements is portrayed.

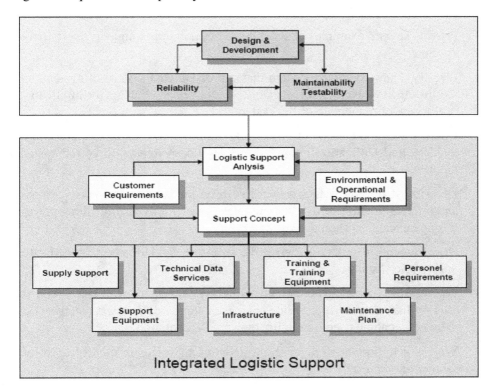

Figure 6-12: Relating ILS to the System Life Cycle
Source: [ASD, 2009]- Aerospace and Defence Industries Association of Europe

The requirements for reliability resulting in the need of maintainability and testability are driving factors. These needs are analyzed in a Logic Support Analysis (LSA) based upon Customer Requirements and the Environmental and Operational Requirements. Consistent with our discussion in Chapter 4 of the importance of driving concepts and principles, a support concept is to be developed that is then utilized in generating the individual ILS supporting systems.

KNOWLEDGE VERIFICATION

1. What is the difference between the management of systems and the leadership of systems?

2. How does the type of as well as the lifetime of a system affect its management and/or leadership?

3. Provide several examples of short longevity and long longevity (sustainable) systems.

4. Identify a large complex system that can be called a system-of-systems that requires the involvement of an extended enterprise. What are the implications of the lack of ownership of the system?

5. Confirm that the generic (T) life-cycle model composed of Definition, Production, and Utilization related stages are applicable to all of the systems identified in (3).

6. Describe situations in which an iterative development, an iterative development and implementation, and an incremental or progressive acquisition approach to stage execution should be utilized.

7. How can requirements and their verification and validation be managed within a life cycle structure?

8. For various types of system elements in a system-of-interest with which you are familiar, identify appropriate sources of standards and best practice documents that can be used in guiding their implementation.

9. Why is it important to involve all actors at all stages in the life cycle of a system-of-interest?

10. What is an Integrated Project Team and what is its purpose?

11. Are life cycle models and the enterprise processes systems? If so, what are the elements of these systems-of-interest?

12. Define a life cycle model as a matrix of Technical Processes with stage outcomes and the basic contributions of each process in each stage for some type of System-of-Interest. (Try making a matrix for the management of a life cycle model System-of-Interest.).

13. Define a set of processes that are appropriate for a type of System-of-Interest. (Continuing with 12, by defining a tailored set of processes that can be applied in managing life cycle model system-of-interest.)

14. What type of documentation elements would be appropriate for life cycle model and process management that result from 12 and 13?

15. Describe the relationship between system life cycles and product life cycles.

16. Identify support requirements for a System-of-Interest with which you are familiar.

Chapter 7
Data, Information and Knowledge

Knowledge and wisdom are, or at least should be,
the ultimate goals of all human activities.

Knowledge that is available to and utilized effectively by competent decision makers provides wisdom that is a basis for continual improvement of enterprise infrastructure system assets, the systems delivered as products and/or services, the formation and deployment of respondent systems as well as the processes and life cycle models that support system related work.

Knowledge can be explicit, that is documented in some manner or tacit (implicit) in which case the knowledge generally comes from an underlying understanding of undocumented *patterns* of system and element relationships. The knowledge may be *abstract* emphasizing concepts and principles or *concrete* in which case the knowledge is related to the details of "what is" or "how to" within the context of the operations of an enterprise grouping (team, board or project). It is important that individuals as well as the group collectively are able to understand concepts and principles as well as relevant concrete details since decisions are made at all levels, by individuals as well as various groups.

In the Knowledge Element of a Change Management System, knowledge is gathered that is to be utilized as feedback in decision-making within the Change Management Element, in Project Management or in line organization management as well as feed forward to enhance capabilities in the "how to" in accomplishing actual changes.

FROM DATA TO WISDOM

The gathering of knowledge occurs in steps. First, the raw *data* that is gathered must be interpreted. It is through interpretation by placing the data into their proper class that *information* is obtained. For example, the data 18.5, 19.6, 20.1, 18.3 only become information when it is known what the data represents. Knowing that the data are temperatures in degrees Centigrade transforms the data into information as portrayed in Figure 7-1. The data can be enhanced and provide new information by measuring the data in respect to another information class, for example, evaluating the data in respect to the temperature bounds of 18 to 22 degrees Centigrade. A Measurement Element determines that the temperatures are within bounds and generates new data, for example (in bounds or out of bounds). Once again, the data (in or out of bounds) is only meaningful when it is known that it belongs to the class called temperature bounds.

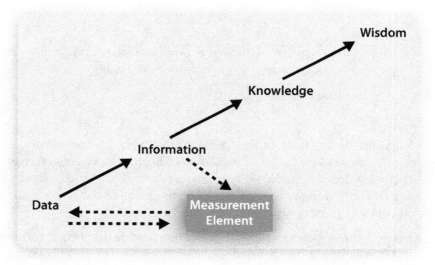

Figure 7-1: Data, Information, Knowledge and Wisdom

The data gathered, including the measured data, and interpretation as information is useful, but they can only be transformed into *knowledge* when related to other information where relationship patterns become evident. For example, knowing that temperature measurements have consistently resulted in (in bounds) coupled together with quantitative information about the design of the climate control system that achieved this behavior provides knowledge about a working solution. Such knowledge can be used in making decisions leading to change and in accomplishing changes in future versions of the climate control system. The transformation of knowledge into effective decision-making demonstrates the final step in Figure 7-1, namely achieving *wisdom*.

The relationship between data, information and knowledge has been stated succinctly by Börje Langefors [Langefors 1973] in the following *infological equation*.

I = i(D, S, t) where:

I is the information achieved by an interpretation process i

D is the data with regard to some previous knowledge S

and t is the time at which the interpretation is made

In this equation S corresponds to the knowledge of an information class.

Consider another example of the data, information and knowledge relationship in respect to Process Capability Assessment as described in Chapter 5. Data for process outcomes is collected as the processes are carried out (executed). The data may be of the type (not performed, performed) for particular outcomes. Until it is known what information this data is related to, it is just data. Knowing that the data is related to a process outcome, such as "requirements validated" converts the data to information. In a process assessment (Measurement Element) this outcome may be assigned a numerical value such as (2). Once again, this generated data becomes information when related to a process maturity class scale resulting in a measurement result like "repeatable". Being able to associate requirement validation information as well as the assessment resulting in data and information concerning other process outcomes builds knowledge about the performance of the process. Wisdom comes from utilizing this information in effective decision-making related to changes directed towards process improvement.

A Broader View of Information

Information is a much wider concept than is implied in the previous discussions about the transformation from data to information. Information is provided in a variety of media. For example, drawings, models, voice recordings and video films all provide information and are important elements in knowledge formation.

A variety of models of various forms have been presented in this book that hopefully have contributed information for the reader in forming knowledge about what it means to "think" and "act" in terms of systems.

DATA AND INFORMATION MANAGEMENT

The wide spread availability of data and information provided via computing and networking resources has revolutionized our way of thinking and working. Prior to the age of computers, data and information was mostly available in paper media.

Since the advent of computing and then networking, several generations of data and information management technologies have led to a data and information explosion. It has been estimated that more data has been generated in the five-year period from 2005 to 2010 then was produced in the previous 40,000 years. [www.c-data.nl] As computing and network technologies have increased in capacity and processing power, the wide scale usage of graphics, recordings and video has also become a vital part of base on information that contributes to knowledge formation.

In his excellent, and early, book on the growing problems with information, Wurman points to the accelerated pace of data and information collection and dissemination due to the developments in Information Technology.

*"Information Anxiety is produced by the ever-widening **gap** between what we understand and what we think we should understand. It is the black hole between data and knowledge, and it happens when information doesn't tell us what we want or need to know."*
Richard Saul Wurman [Wurman, 1989]

The availability of too much information causes anxiety for individuals as well as for groups. The relevance of, presentation of as well as the quantity and the quality of information are all important knowledge enablers. In operating an enterprise, it is vital to gain control over the information anxiety problem. Based upon Wurman's definition, we could perhaps define information quality as being related to how much the **gap** can be reduced. The true measure of the quality of information in an enterprise context is how well information of the appropriate quantity and categories is gathered, retained and utilized in knowledge building and decision-making.

Information quantity and quality is a growing concern in the life cycle management of complex systems. In particular, a significant factor is the increased outsourcing of product and service creation, operation and maintenance in large supply chains as discussed in Chapter 4 and as illustrated in Figure 4-10. The problems are not only related to the volume of information but are related to incompatibilities of data representation and information classification. These problems are compounded in international enterprises where various language and cultural differences exist. Thus, it is vital that information is classified and gathered in a manner that is understandable and useful for the parties that have the need to know.

The ISO/IEC 15288 standard provides a process for Information Management as well as for Measurement that can be tailored to meet data and information needs in the life cycle management of systems.

Information Management	provide relevant, timely, complete, valid and, if required, confidential information to designated parties during and, as appropriate, after the system life cycle
Measurement	collect, analyze, and report data relating to the products developed and processes implemented within the organization, to support effective management of the processes, and to objectively demonstrate the quality of the products

Classifying Information

In order to provide a basis for common understanding amongst people and the computer software agents that collect, store and retrieve information, the classification of information is a perquisite. As computer technology has advanced from sequential processing of files of data to randomly accessed data bases and now to interactive web access various approaches to describing information structure and content have arisen. The development of descriptions of information reflects the needs of the system and is based upon the identification of concepts, classes and properties of the data to be processed.

Wurman [Wurman, 1989] identifies five fundamental ways of organizing information; namely by *alphabet*, by *category*, by *time*, by *location*, or by *continuum*. These modes of classification are applicable to almost any endeavor – from personal file cabinets to the operation of multinational corporations. As an example consider classifying information for the automotive industry where the name of the company is used for an *alphabet* classification, the identification of auto models for *category* classification, the year of production for *time* classification, the place of manufacture for the *location* classification, and consumer report ratings for *continuum* classification. The following summarizes the general usage of the Wurman categories.

Alphabet – A natural means of classification for western languages. Lends itself nicely to organizing large quantities of information.

Category – Typically used to categorize products and can also be used to categorize services. Lends itself well to organizing items of similar importance.

Time – Works best as an organizing principle for events that happen over fixed duration. Time is an easily understandable framework from which changes can be observed and comparisons made.

Location – A natural form to choose when trying to examine and compare information that comes from diverse sources or locales.

Continuum – Used for assigning values or weights to information. Organizes items by magnitude from small to large, least expensive to most expensive, by order of importance, etc.

The reader certainly already consciously or unconsciously deploys these modes of classification in his/her daily life. These categories are generic and can be related to any class of information. Most essential is identifying the concepts under which information is organized in a rational manner thus providing a basis for common understanding as well as guidance for the software agents needed to collect, store, reference and present the data and information in a palatable manner.

Taxonomies and Ontologies

While the classification of information has a long history, it is with the wide scale introduction of IT (Information Technology) that significant discussion and even much "hype" surrounding information classification have arisen. In this section, we consider the main ideas of taxonomies and ontologies. There is significant literature available on these and related topics that should be investigated by the reader via a web search. Consider the following two slightly reduced definitions taken from Wikepedia:

> **Taxonomy** is the practice and science of classification. The word finds its roots in the Greek, *taxis* (meaning 'order', 'arrangement') and *nomos* ('law' or 'science').

> **Ontology** (from Greek) is the philosophical study of the nature of being, existence or reality in general, as well as the basic categories of being and their relations.

Many cite the work of Aristotle in what he called *metaphysics* as the beginning point for the establishment of ontologies. His work in multiple disciplines as described in Chapter 1 drove him to see the need for unification by classifications and relationships. Through the centuries there have been many significant works in classification, for example, the enormous effort by the 18th century Swedish scientist Carl von Linne in creating biological taxonomies for species of plants and animals.

In practice a taxonomy can be seen as a restricted form of ontology where things are simply categorized in a *"is a"* relationship indicating that an object is a member of a class of objects. Further relationships are not described. Taxonomies are typically organized into a hierarchy of classes reflecting the concepts utilized in formalizing the classes. Within a class the taxonomy may well be organized according the to information categories that Wurman identifies; that is by *alphabet*, by *category*, by *time*, by *location*, or by *continuum* or combinations thereof.

Generally, we can state that ontology is the theory of objects and their relationships. An ontology provides criteria for distinguishing various types of objects (concrete and abstract, existent and non-existent, real and ideal, independent and dependent) and their ties (relations, dependences and predication). Based upon what has been covered in this book, we can see a strong relationship between the notion of ontology and system. On the practical side, ontologies (and taxonomies as well) provide a means of forming an information model in an area of discourse or even for a specific system-of-interest.

The OWL Semantic Web technology [Schreiber and Wood, 2004] often referred to as Web 3.0 is based upon the usage of ontologies. The treatment of ontologies as systems in the context of this technology is described in the case study interlude following this chapter.

In the broader interpretation of information presented above, it was noted that as various computer-based multimedia technologies have evolved. Thus, the treatment of voice, pictures, drawings, films, etc. has radically extended the types of "data" that can be processed within the information technologies. Files containing these media representations become fundamental data that must also be classified as a part of information models.

The kaleidoscope view of systems described earlier in the book is evident even when considering the creation and usage of data and information. In fact, life cycle information for a system as well as its instances can also be treated as a system composed of information elements. Further, various kaleidoscope views of a system will result in various sets of information each of which, to relevant parties can be viewed as "their" system. Understanding the structure and organization of system life cycle related information permits the extraction of value and significance. It is an absolute information quality requirement and forms the basis for prudent decision-making.

Gathering Data, Information and Knowledge

During the life cycle of the sustained value-added and infrastructure system assets that are managed by an enterprise, data, information and knowledge is gathered in the manner described in the previous sections of this chapter. In respect to the fundamental life cycle transformations identified in Chapter 4, the collection of data, interpretation as information, potential measurement and assimilation as knowledge is portrayed in Figure 7-2.

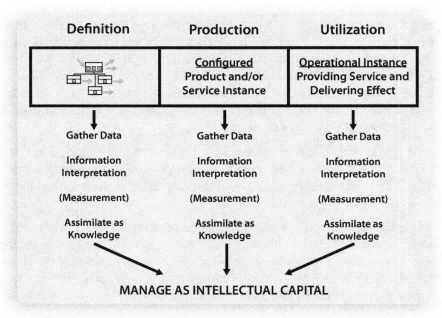

Figure 7-2: Assimilating Knowledge as Intellectual Capital

The data, information and knowledge gained during the life cycle are of various types. It can be related to system definition, to the production of instances as products or services or to the services provided during utilization. It can also be process knowledge related to the performance of the processes leading to the system definition and/or to a product or service in various life cycle stages. A prerequisite is of course the definition of an information model for the system.

Building Information Models

In building the information model, it is useful to use as concepts the terminology established by the ISO/IEC 15288 as a basis for classification. It could be done according to the data and information generated from processes that are to be implemented for the various stages of the life cycle; it could be also more generally based only upon the stages. An illustrative information model classification is portrayed in Figure 7-3. In this model a hierarchy of information classes is established where the leaves (end nodes) of the hierarchy identify classes into which data and information is to be gathered. Do not forget that the information may well be based upon models (textual or graphic) and can include voice and video media.

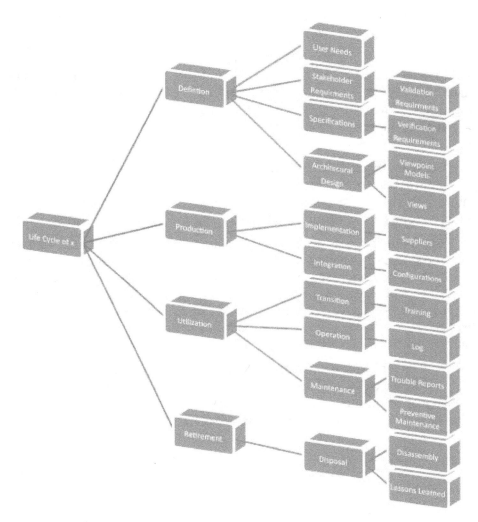

Figure 7-3: An Illustrative Life Cycle Information Model

The figure illustrates data and information collected for technical processes; however it can and should be extended with data and information related to the other ISO/IEC 15288 standard categories; namely Agreements, Organizational Project-Enabling and Project. The collections of data, information, measurements and knowledge associated with baselines and configurations as described in Chapter 4 are very germane and must be taken into account in organizing the data and information. Recall that baselines can be associated with project information (scope, cost, schedule) or system product and/or service. Furthermore, some baselines actually represent a configuration in respect to descriptions and or product/ service instances. As systems progress through the establishment of baselines and the creation of configurations, the knowledge gathered about the system, product, service, and processes is vital in the decision making of the CCB as well as in providing guidance in making changes.

As indicated in Figures 7-2 and 7-3, the assimilated knowledge becomes part of the intellectual capital of the enterprise, its groups, teams and individuals. Thus, enterprises are wise to formalize the structure of and the processes related to the gathering of knowledge in a Knowledge Management System. Naturally, this system like all others has a life cycle that must be managed [Herald, Berkemeyer, and Lawson, 2004].

Within the domain of a Knowledge Management System, the Life Cycle Models as well as the sets of Processes that are used in the management of Systems-of-Interest from the system portfolio is true enterprise intellectual capital. They must be captured in the information model of the Knowledge Management System.

Equally important is the knowledge attained from structuring and analyzing thematic systems in studying how to address enterprise problems and/or opportunities. Thus, the approaches to Systems Thinking provided by Senge, Boardman, Checkland and others form a basis for knowledge assimilation and the models created must become part of the information base. It is wise to collect "lessons learned" from such studies so that the knowledge is truly captured as part of the enterprise intellectual capital.

With intellectual capital that assists in understanding the dynamic operations of systems as well as in understanding how to effectively structure and operate change management and life cycle management, an enterprise is well equipped with the wisdom required to "Think" and "Act" in terms of systems.

CREATIVE THINKING

"Imagination is more important than knowledge"
Attributed to Albert Einstein

In pursuing new avenues in response to problems and opportunities or in uncovering new problems or opportunities the enterprise (including its CCB members) need to develop the capability to think in a creative manner. The approaches to Systems Thinking presented in Chapter 2, along with all of the available literature related to this topic provide knowledge inputs in this regard. Creative Thinking is certainly related to Systems Thinking but there are some aspects of thinking creatively that are not often discussed within the scope of Systems Thinking. Several renowned gurus have developed approaches to Creative Thinking and a full treatment is obviously beyond the scope of this book. Again, a web search is recommended.

Synectics

One approach that has been successfully applied by enterprises and individuals is *Synectics* developed by the late MIT professor W.J.J. Gordon [Gordon, 1961]. Gordon identified the following forms of analogies to be utilized in problem solving and opportunity evaluation situations:

personal analogy – one identifies with the elements of the problem and plays the roll of key elements.

direct analogy – one compares parallel facts, knowledge, or technology. Comparisons with biological and physical systems are often rewarding in this regard.

symbolic analogy – one uses objective and impersonal images to describe the problem in a technologically inaccurate but esthetically satisfying manner.

fantasy analogy – one imagines the best of all possible worlds where everything is possible.

In using a *personal analogy*, an individual can imagine herself/himself as an actor in the system. For example, one can play the role of a control element, controlled element or measurement element. "If I were the computer operating system what path of action would I take in this situation?" This is an analogy that can be of use in attempting to understand or debug an important software element.

The use of *direct analogy* is the one of most central foundations of Systems Thinking. It was Ludwig von Bertalanffy that in the late 1920s pointed out the analogies between biological systems and other physical and eventually non-physical systems. This is evident in the usage of cybernetics as a control mechanism for physical as well as organizational systems. Seeing the similarities by comparing facts, knowledge or technology provides insight. Certainly, the ISO/IEC 15288 provides an instrument to support direct analogies by providing concepts and principles as well as processes that can be utilized in a common manner for the life cycle management of all types of man-made systems.

A *symbolic analogy* or *fantasy analogy* can be used to gain perspective to a real problem or opportunity. Looking at the problem from another angle in which characterization of the problem or opportunity is expressed in some, perhaps unconventional manner or based upon unrealistic assumptions can provide insight in identifying new approaches. Such analysis can lead to the discovery of paradoxes as described in Chapter 2.

Concept Maps

The identification of concepts for a system is a creative activity that requires knowledge and experience as well as imagination. This was pointed to as a success factor in Chapter 4 and illustrated in the interlude case study following that chapter. As described above, the identification of concepts is also a perquisite to building an information model for an area of discourse or for a system-of-interest. One method supported by a highly useful tool is that of concept maps. [Novak and Cañas, 2008] There are several examples of utilizing concepts in their technical report as well as information concerning the Cmap tool on the website www.ihmc.us.

Novak and Cañas characterize concept maps as graphical tools for organizing and representing knowledge. They define, in manner consistent with the presentation in Chapter 4, a concept as a perceived regularity in events or objects, or records of events or objects, designated by a label. Another characteristic of concept maps is that the concepts are represented in a hierarchical fashion with the most inclusive, most general concepts at the top of the map and the more specific, less general concepts arranged hierarchically below. The hierarchical structure for a particular domain of knowledge also depends on the context in which that knowledge is being applied or considered. It is interesting to note that they choose to describe the features of concept maps as the concept map portrayed in Figure 7-4. The reader is encouraged to explore this model and relate it to personal experiences with concepts.

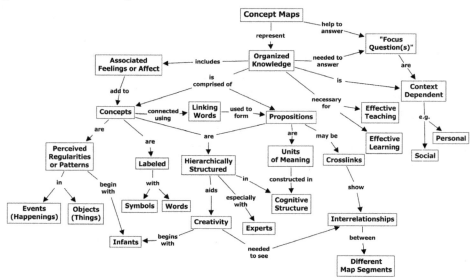

Figure 7-4: The Features of Concept Maps as a Concept Map

In summary, creative thinking approaches like analogical reasoning and/or concept maps in supporting and organizing thinking provide useful approaches to be exploited by individuals and groups in the acquisition of knowledge.

THE LEARNING ORGANIZATION

In this book, approaches to achieving the capability to "Think" and "Act" in terms of systems have been presented. The bottom line, of course, is how well individuals of an organization as well as its organizational elements (enterprise, groups in the form of projects or task forces and change control boards) assimilate this knowledge and utilize the gained wisdom in improving their individual and collective capabilities.

Peter Senge in his classic Fifth Discipline book describes the ingredients that are necessary to achieve a learning organization. Senge sees Systemic Thinking as the discipline that integrates a set of related disciplines; namely:

- Personal Mastery
- Mental Models
- Shared Vision
- Team Learning

A detailed description of the theory and practice of these disciplines can be found both in Senge's book "The Fifth Discipline" [Senge, 1990] as well as in "The Fifth Discipline: Fieldbook" [Senge, et al, 1994]. The reader is encouraged to consult these sources for a thorough explanation of the disciplines. In particular, the Fieldbook provides a large number of practical examples of how the disciplines have been applied through the use of a variety of methods and tools. The following summarizes the main aspects of the disciplines and how they are related.

Personal Mastery

Naturally, the starting point for all learning is personal. It is the pre-condition as indicated in the following quotation by Senge:

> *"Organizations learn only through individuals who learn. Individual learning does not guarantee organizational learning. But without it no organizational learning occurs."*

Personal mastery is likened to a journey whereby a person continually clarifies and deepens personal vision, focuses energy upon it, develops patience in seeking it, and in this way apparently increasingly views reality in an objective manner. Personal mastery cannot be forced upon people; however, it is a part of organizational strategy that must be explained and encouraged. There are sources of tension and conflict that most likely will occur during the journey in respect to the gap between personal vision and reality.

Creative tension is positive and arises from a commitment to change where the person experiences energy and enthusiasm. Current reality is moved towards personal vision.

Emotional tension is negative and arises from a lack of belief that change is possible. This can lead to feelings and emotions associated with anxiety and pulls back personal vision towards reality.

Structural conflict may arise when there is a disbelief in one's own ability to fulfill intrinsic desires and rests upon assumptions of powerlessness or unworthiness. A lack of belief in oneself threatens to erode personal vision.

Systems Thinking contributes to Personal Mastery by helping the individual to explain the dynamic nature of structures in their lives (in personal as well as in work related situations).

Mental Models

Mental models are conceptual structures in the mind that drive cognitive processes of understanding. They most often invisibly define an individual's relationship with other individuals and with the world in general.

The discipline of mental models aims to train people to appreciate that mental models do indeed occupy their minds and shapes their actions. In managing mental models, the individual must develop skills of *reflection* and *inquiry*, for example, bringing assumptions of mental models to the surface and testing adequacy via inquiry and reflecting upon their implications. In a learning organization, people are able to test many mental models – their own as well as those of colleagues. A facilitative organizational structure is required to promote these skills. Note that the various forms of analogical reasoning as well as the "language" of Systems Thinking can be of assistance in formulating mental models.

Shared Vision

In contrast to personal visions, shared vision is related to the mental pictures of people throughout the organization. Shared vision refers to shared operating values, a common sense of purpose, and a basic level of mutuality. It extends personal mastery into a world of collective aspiration and shared commitment. Shared vision provides a focus and energy for learning (resulting in negative entropy as described in Chapter 5). Senge describes this as generative learning as opposed to adaptive learning that corresponds to the notions of pro-active and reactive decision-making also described in Chapter 5. By being generative the organization expands its capacity to create its own future (pro-active) rather than be created by events of the moment (reactive).

Shared vision fosters risk taking and experimentation. It engenders leaders with a sense of vision who wish to communicate the visions to inspire others to share and assimilate as their personal vision – the art of visionary leadership. Such leadership may come top down in an organization or bubble upwards. In either case a truly shared vision takes time to emerge and shape how things are done.

Systems Thinking explains the spread of shared vision in generative learning as a reinforcing (growth) process. The communication of ideas gathers pace and the vision becomes increasingly clear, leading to rising enthusiasm. The growth process can be counterbalanced by a (limiting) factor. For example, when too many people get involved. The more people the greater the potential for diversity of views to break out. Another limiting factor arises when people see the gap between shared vision and current reality leading to negative feelings and erosion of the shared vision goals.

Team Learning

The aim of team learning is to achieve alignment in people's thoughts and energies. A common direction creates a feeling that the whole team achieves more than the sum of its team members. If people are not aligned, important qualities of the learning organization such as empowerment may actually increase conflict. Empowerment involves the assignment of authority and responsibility to individuals or groups.

Successful team learning is related to achieving a balance between *discussion* and *dialogue*. Discussion is the communicative means where different views are presented and defended in the search for an optimal view to support a decision that must be made. Dialogue is communication of a different nature where people suspend their views and enter into deep listening in the sense that the listener visits and explores the mental models of other team members. The listener attempts to see through the eyes of other team members.

The perspectives and tools of Systems Thinking are vital for team learning. For example, in management teams like the CCB, where dealing with complexities is a primary task. Systems Thinking provides a language through various approaches such as system archetypes, influence diagrams, systemigrams and rich pictures that assist in coming to grips with dynamic complexity. It helps to bring together people's mental models into a shared vision thus generating team learning and understanding as well as a shared sense of purpose.

Systems Thinking

As the reader will observe, it is Systems Thinking that plays the coordinating role amongst the other four disciplines. It provides a combined theory and practices that are essential for the learning organization. In summary, the Systems Thinking contribution to the other disciplines is as follows:

Personal Mastery – helps in continually seeing system interconnectedness as well as the interdependencies between our actions and our reality.

Mental Models – exposes assumptions and tests if these are fundamentally flawed, for instance, by identifying feedback not previously accounted for.

Shared Vision – clarifies how vision radiates through collective feedback processes and fades through conflicting feedback processes.

Team Learning – identifies positive and negative synergy in discussion and dialogue, respectively, the whole becomes more than the sum of its parts.

Building a Learning Organization

Building a learning organization requires a dedication to purpose. It does not happen overnight and it requires continual encouragement in respect to all five disciplines in order to become reality. It can be a costly process. However, when successful, the ROI (Return on Investment) can more than justify the cost both in tangible and intangible positive effects. There can be a variety of reasons for wanting to build a learning organization:

- Because we want superior performance
- To improve quality
- For customers
- For competitive advantage
- For an energized, committed work force
- To manage change
- For the truth
- Because the times demand it
- Because we recognize our interdependence
- Because we want it

Whatever the reason or reasons, it is essential to establish an organizational architecture that provides the structures from which positive learning behaviors develop and intellectual capital is increased. Senge presents such an organizational architecture and its relationship to the reinforcing (growth) loop that provides the basis for achieving a learning organization environment as portrayed in Figure 7-5.

In formulating an organizational architecture, the organization must establish and continually review their guiding ideas in the form of abstract concepts and principles as well as policies. A willingness to innovate in respect to sustained infrastructure system assets and to make explicit the underlying theories, methods and tools are also essential elements.

The organizational architecture forms the framework upon which the skills and capabilities of individuals and organizational elements are continually improved. This leads to increased awareness and sensibilities and then to strengthening of attitudes and beliefs. With the proper energy (negative entropy), this growth cycle can continue to result in improvements in the organizations ability to achieve its purpose, attain its goals, and accomplish its missions. True wisdom has been achieved.

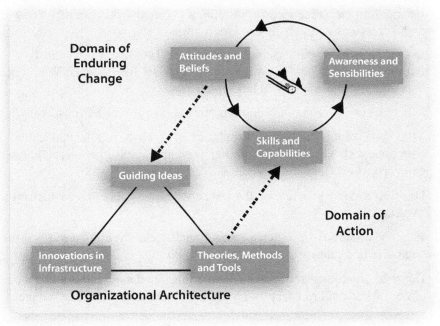

Figure 7-5: Organizational Architecture and Enduring Change

KNOWLEDGE VERIFICATION

1. What is the relationship between data, information, measurement, knowledge, and wisdom?

2. For various types of sustained systems create examples of data, information, measurement and knowledge in respect to system definition, production and consumption.

3. Explore the data produced in Consumer Satisfaction Index investigations. Describe how this data becomes information and how it can contribute to knowledge and wise decision-making.

4. Use at least three of the *alphabet*, *category*, *time*, *location*, and *continuum* information classifications in building a taxonomy for a family of products and/or services with which you are familiar.

5. Discuss how well the terminology from the ISO/IEC 15288 can be used in building the concepts for an information model for systems. What types of data and information should be collected.

6. What role can baselines and configurations play in contributing to knowledge?

7. What types of information knowledge are used as the intellectual capital of an enterprise with which you are familiar? And how is information and knowledge categorized into an ontology?

8. Develop a concept map for the concepts of a system with which you are familiar.

9. Identify situations in which personal analogies, direct analogies, symbolic analogies and fantasy analogies can be applied.

10. Describe situations from personal experience that can be related to the disciplines of Personal Mastery and Mental Models as described by Senge.

11. What problems are encountered in an organization in building a Shared Vision? Describe some personal experiences in this regard.

12. Why are both discussion and dialogue so important for Team Learning?

13. How is Systems Thinking deployed as a unifying discipline for the other four disciplines of the learning organization?

14. What problems would be encountered in building a learning organization in an enterprise environment with which you are familiar?

Interlude 4: Case Study in Ontology Life Cycle Management

This case study provided by Marie Gustafsson is taken from a published article entitled: "Ontology Development and Deployment using ISO/IEC 15288" [Gustafsson, 2006]. The publication was based upon a project assignment performed by Marie at a course delivered by your author at the University of Skövde, in 2005. This project definitely provides evidence that systems have a broad definition. In this case ontologies are viewed as systems that can be and should be life-cycled managed. The work performed also became part of Marie Gustafsson's doctoral thesis [Gustafsson, 2009] and was supported by the Swedish Agency for Innovation Systems.

Abstract

We propose that ontologies can be regarded as systems, in that they can be seen as descriptions, agreements, and products. In viewing ontologies as systems, it is demonstrated how the ISO/IEC 15288 system life cycle processes can be applied to ontology development and deployment. A short case study illustrates how these processes can be applied in creating an ontology for oral medicine. An advantage of applying this standard is a beneficial level of adaptability to different ontology development configurations. Another is the standard's consideration for non-technical elements, which is relevant in attempting to transfer the conceptualizations of a group into a technical specification.

INTRODUCTION

It has been widely suggested that ontologies can be used for sharing a common understanding of the structure of information among people and software agents. Several methodologies for developing ontologies have been proposed, but ontology construction largely remains a craft rather than an understood engineering process. The lack of actual deployment of ontologies, whether in the form of applications built on ontologies or as published ontologies available for reuse, is a major problem.

Proposed methodologies for ontology engineering have drawn largely on terminology from software engineering. [Uschold and King, 1995], [Grüninger and Fox, 1995] and [Fernández-López, et al., 2000] The ISO/IEC 15288 standard, on the other hand, equips us with a domain independent way of understanding the nature of and composition of man-made systems, as well as their movement through life cycles. It is required by the standard that a life cycle model be developed for each system of interest to which the standard is to be applied. In viewing ontologies as systems, techniques from systems thinking and systems engineering can be applied to structure the development and deployment of ontologies. These techniques also aid in change management and in enabling us to integrate both technical and non-technical system elements. Inspired by existing methodologies, we propose an approach for developing ontologies based on the ISO/IEC 15288 standard.

If we look at one of the most common definitions of ontology, Gruber's "explicit specification of a conceptualization" [Gruber, 1993], it is interesting to note what he thinks today: "In retrospect, I would not change the definition but I would try to emphasize that we design ontologies. The consequence of that view is that we can apply engineering discipline in their design and evaluation" [Lytras, 2004]. We believe that using the ISO/IEC 15288 standard can add valuable engineering experience to the design of ontologies. We begin with an overview of ISO/IEC 15288 and describe how ontologies can be seen as systems. After an overview of other work on ontology development, we describe how the ISO/IEC 15288 system life cycle processes can be applied to ontology development and deployment. A short case study is followed by a discussion of the advantages of using this approach.

ISO/IEC 15288 SYSTEM LIFE CYCLE PROCESSES

ISO/IEC 15288 System Life Cycle Processes has as its primary goal to "provide a basis for international trade in system products and services". To meet this goal, the standard provides guidance for defining structures and boundaries of systems, structuring life cycles for systems, as well as processes for enterprise management of systems, for acquisition and supply agreements, for project management of system related work, and for carrying out the technical aspects of systems. Its creators hold that "the standard can be applied to any type of man-made system."

A system, according to the standard, satisfies a need, yields a potential set of services, and when used in an operational context, provides effects. A system-of-interest is the system on which focus is placed. The system-of-interest is made up of other systems and system elements, which provide services to the system-of-interest. The development of a life cycle model is required, but the standard does not require a specific model. However, as guidance a description of a common life cycle model structure is presented, constituted by the concept stage, the development stage, the production stage, the parallel stages of utilization and support, and finally the retirement stage. It is not to be seen as a sequence of stages; iterating between the stages is possible and often necessary. The standard provides a recursive system composition, and may be reapplied for systems of any type, no matter where the system is placed in a system hierarchy. Also, enabling systems are identified as systems necessary to the system-of-interest's progress through its life cycle, but which are auxiliary to its purpose. Further, a set of processes is provided that can be applied and tailored to be used at different stages of the life cycle. These processes can be divided into categories of Enterprise Processes, Agreement Processes, Project Processes, and Technical Processes. More details on these processes can be seen in the section on using ISO/IEC 15288 for ontology development.

ONTOLOGIES AS SYSTEMS

Ontologies can be seen as systems in that they are agreements – which can be seen as a kind of system – of what we want to be able to describe. An ontology can also be seen as a product, a designed entity. If "systems exist only as descriptions" is applied to ontologies, we can say that an ontology is essentially just a description: It is a description of what a group of people or an organization feel they need to be able to describe, discuss, and share information about.

As stated previously, systems satisfy needs, and the needs to be fulfilled by an ontology can be to [Noy and McGuinness, 2001]: share a common understanding of the structure of information among people and software agents; enable reuse of domain knowledge; make domain assumptions explicit; separate domain knowledge from the operational knowledge; and analyze domain knowledge. Effects and potential services of an ontology relate to these reasons.

A system element is either an ontology (composed of classes, properties, and restrictions), which is not further decomposable, or it is another ontology, which we are reusing. Typically, an ontology consists of both classes and relations defined for just this ontology, as well as reuse of other ontologies. Apart from being a system element in another ontology, an ontology can also be a system element in an ontology-based application.

Another important aspect of systems is that they can be divided into systems-of- interest. For ontologies, several different types of ontologies are often distinguished, mainly [Pinto and Martins, 2004]: representational ontologies, upper-level ontologies, domain ontologies, and application ontologies.

Different types of ontologies, forming a hierarchy from general to more specific, can be said to be systems-of-interest at different levels. Though this paper will take the system-of-interest to be the ontology, the ontology could also be seen as a system element in an ontology-based software application. We could thus apply ISO/IEC 15288 to the level of the ontology-based software development as well.

Stakeholders in an ontology are both humans and machines, and there are needs to be fulfilled to accommodate both. Classes of human users of ontologies are: developers of large scale reference ontologies; users who need to augment or adjust existing ontologies for a particular application; users who need to construct small, specialist ontologies for a particular use; and applications developers who want to use ontologies.

There are both suppliers and consumers of ontologies, and a certain ontology may both consumed (reused) and be supplier for (be available for reuse to) other ontologies.

In viewing ontologies as systems, we can apply techniques from systems thinking and systems engineering to structure the development and deployment of ontologies, and to aid in change management. Before moving on to a description of how ISO/IEC 15288 can be applied to ontology development, some related work in this area is described.

ONTOLOGY DEVELOPMENT

Several methodologies for developing ontologies have been proposed, a few of which will be described below. Some topics of special interest in ontology development – reuse, evaluation, and maintenance – will also be briefly discussed.

Methodologies for Ontology Development

Terminology in ontology engineering draws on software engineering, where the usually accepted stages through which the ontology is built are specification, conceptualization, formalization, implementation, and maintenance [Pinto and Martins, 2004]. The activities of knowledge acquisition, evaluation, and documentation should be performed during the whole life cycle. A difference is that in software engineering knowledge acquisition is rarely present, while it is central to the ontology building process. Another difference is the separation of conceptualization and formalization. Uschold and King's method [Uschold and King, 1995] starts with the identification of purpose and scope. The key concepts and relationships of the domain are then identified and captured in unambiguous textual form, and then mapped onto a precise terminology. Thereafter, they are coded in a formal knowledge representation language and appropriate knowledge is reused from existing ontologies. When the ontology is integrated, it is evaluated and documented for later reuse and modification.

Grüninger and Fox [Grüninger and Fox, 1995] propose a methodology based on development of knowledge base systems (KBS) using first order logic. Like Uschold and King's approach, Grüninger and Fox propose that one begin with capturing motivating scenarios. Then, informal competency questions are used to determine the scope of the ontology. These questions and their answers are used to extract main concepts and their properties, relations, and axioms, which are first semi-formally specified, and then given a formal specification. Finally, the competency questions are used to evaluate the system.

The METHONTOLOGY approach [Fernández-López, et.al., 2000] can be used to build ontologies from scratch, reuse other ontologies as they are, or for reengineering them. It includes an ontology development process, a life cycle based on evolving prototypes, and particular techniques for carrying out each activity. The development process consists of: scheduling, control, quality assurance, specification, knowledge acquisition, conceptualization, integration, formalization, implementation, evaluation, maintenance, documentation and configuration management. The life cycle identifies stages through which the ontology passes and where there are interdependencies with the life cycles of other ontologies.

Despite these methodologies, it is often observed that ontology building remains a craft rather than an understood engineering process [Gómez-Pérez and Rojas-Amaya, 1999]. It has been noted that there is not enough support for the ontological engineer in making design choices at different levels. A refinement of these guidelines is needed [Jones, et.al., 1998]. While many methodologies give general guidelines for each stage, there is a lack of information about how to accomplish these, for example what actions and decisions need to be taken at each stage [Valarakos, et.al., 2005]. One of a few exceptions to this is the METHONTOLOGY approach [Fernández-López, et.al., 2000], where more explicit guidelines are provided.

Ontology Reuse

Ontology reuse is a process in which available ontologies are used to generate new ontologies. Two reuse processes can be identified, fusion and composition [Pinto and Martins, 2004]. When fusion is used, an ontology is built by unifying two or more different ontologies on the same subject. With composition, an ontology is built in one subject reusing one or more ontologies in different subjects. Source ontologies are aggregated, combined, assembled together, to form the resulting ontology. While existing ontology building methodologies recognize reuse as part of the overall ontology building process, they do not explicitly address the issue. There are still not many methodologies that support ontology building through the full range of reuse, including both fusion and composition [Pinto and Martins, 2004].

Ontology Evaluation

Evaluation is an important part of the ontology development process. In categorizing evaluation approaches [Brank, et.al., 2005] it is found that the ontology is either: compared to a "golden standard"; used in an application of which the results are evaluated; compared with a source of data relevant to the domain; or evaluated by humans assessing it according to pre-defined criteria. Evaluation in ontology development can also be divided into technical and user assessment [Pinto and Martins, 2004]. In the technical evaluation, the ontology is judged against a framework, where there are two activities: verification (correctness according to the accepted understanding of the domain) and validation (correspondence with what it is supposed to be, according to the specification requirements document). In user assessment, the usability and usefulness of the ontology and its documentation when (re)used or shared in applications is evaluated from the user's point of view.

Ontology Maintenance

Knowledge models will almost inescapably change during the process of building and using a knowledge-based system. The formulation of the model might lead the expert to revise it, and more so when feedback is given from applying the model in a real or simulated world. Another factor that may lead to change is that finding requirements is problematic, since potential users have difficulties assessing the benefits or possible usages of the new system, and the system itself modifies the work processes when installed. The assumptions on which the model is based can be wrong, originating partly in the domain experts' difficulties in exposing their daily practice.

USING ISO/IEC 15288 FOR ONTOLOGY DEVELOPMENT AND DEPLOYMENT

The standard requires that purpose and outcome be defined for each stage of the life cycle. Life cycle processes and activities are chosen, tailored, and employed in a stage in order to fulfill its purpose and outcomes. The twenty-five system life cycle processes described by the standard are divided into enterprise processes, agreement processes, project processes, and technical processes. For each process, the standard describes a purpose, outcomes of a successful implementation, and associated activities.

We will now look at how different processes can be used during the ontology development and deployment life cycle. For each stage of the system life cycle, relevant processes from the standard have been selected. Most of the processes selected below are from the category of technical processes. The actions suggested for these processes are in part proposed by the standard and in part inspired by earlier proposed methodologies and general research on ontologies. Though a feasibility stage is not included in the illustrative life cycle model described in the standard, it can be added to the model when, for example, the required system services can be designed and developed in several different ways, which is the case for ontology development.

It should be noted that the component processes of this life cycle depends on what kind of ontology is to be developed. Ontologies that are to be large-scale ontologies used by many users would have a different focus and higher needs in rigorosity than those developed to be used by smaller user groups. Also, the manner of construction, for example manual or semi-automatic, also affects what actions should be carried out in the different processes.

1. Concept stage

   Stakeholder Requirements Definition Process: Identify stakeholders, such as domain experts, ontology developers, maintainers, and users. Capture motivating scenarios. Use these to identify the purpose and scope of the ontology and to enumerate important terms.

2. Feasibility stage

   Stakeholder Requirements Definition Process: Refine motivating scenarios, purpose and scope, and important terms from the Concept stage.

   Acquisition Process: Identify possible ontologies for reuse, using the list of important terms as a starting point; evaluate the identified ontologies according to criteria defined in the Stakeholder Requirements Definition Process. Decide whether any of the external ontologies fulfill the particular need, and to what extent internal adaptations are needed. Also, consider if there are sources suitable for semi-automatic ontology construction, which might be used to bootstrap the ontology.

3. Development stage – In this stage it is decided what classes, properties, constraints, and instances should be included in our ontology. This is done informally, that is, not in a representation language.

Acquisition Process: For the ontologies selected for reuse, carry out a more detailed analysis of what concepts are to be reused and what concepts need to be represented in the internal ontology.

Architectural Design Process: Both manual and semi-automatic ontology learning should be considered for defining the architecture of the ontology. Define the classes and the class hierarchy. Possible approaches for developing a class hierarchy are top-down, bottom-up, or a combination of the two. Define the properties of the classes, which describe the internal structure of concepts. Define constraints, which describe or limit the set of possible values of properties. The constraints can deal with cardinality, value type, as well as domain and range. Identify instances of the classes. Defining an individual instance of a class requires choosing a class, creating an individual instance of that class, and assigning properties. Consider using ontology design patterns [Svatek, 2004], to help solve design problems for the domain classes and properties that make up the ontology.

Implementation Process: Decide upon how to encode the classes, properties, and constraints identified in the Architectural Design Process. For example, should the Web Ontology Language (OWL) [W3C Recommendation, 2004a] be used? Along with this decision comes the level of formality aimed for in the ontology.

Verification Process: Assemble reasoning test cases to be used for catching errors when the ontology is encoded.

Validation Process: Check identified ontology components with the stakeholders, especially the domain experts.

4. Production stage – This stage entails encoding the classes, properties, constraints, and instances identified in the Development stage using an ontology representation language. The result is more formal than that of the Development stage, but the degree of detail and formality will still vary depending on the purpose of the ontology.

Implementation Process: Encode the classes, properties, and constraints, identified in the Development stage, in the representational form decided upon. In this process, identified best practices should be used. For OWL, see for example the W3C Semantic Web Best Practices and Deployment Working Group [Schreiber and Wood, 2004].

Verification Process: Check that the encoding corresponds to the specification requirements document. Also, reasoning and test cases identified in the Development stage could be used to identify problems and inconsistencies.

Validation Process: Can the constructed ontology be used in satisfying the motivating scenarios? Check correctness according to the accepted understanding of the domain.

Maintenance Process: Establish criteria for further changes to the ontology and adapt an ontology versioning policy.

5. Utilization stage

Operation Process: Use the ontology as part of an ontology-supported application. This could for example be done using the concepts discussed by Knublauch [Knublauch, 2004], where it is proposed that Semantic Web applications should consist of two separate but linked layers: the Semantic Web Layer makes ontologies and interfaces available to the public, while the Internal Layer consists of the control and reasoning mechanism.

Supply Process: Make the ontology available and known to others by, for example, adding it to ontology repositories.

6. Support stage

Maintenance Process: Make adjustments to the ontology when possible errors surface, or to accommodate changes in domain knowledge, according to the versioning policy.

7. Retirement stage

Disposal Process: When the ontology is no longer applicable or needed, when the changes that need to be made are very large, or when another ontology fulfills the same needs better, it is possible that the ontology should be retired. This entails creating a retirement strategy to address issues of how to map current usages of the ontology to a newer one, among others.

Many of these stages have an iterative nature. For example, work in informally defining classes may bring about changes in the requirements definition, and in formally defining classes we might realize that changes need to be made to the informal definitions. Stakeholder requirements are allocated to different stages. The system lifetimes of ontologies will vary depending on how high up in the hierarchy of ontologies an ontology is and on how extensively it is used.

For ontology construction there are also enabling systems such as ontology editors (e.g., Protégé[1]) and visualizers (e.g., IsaViz[2]). For constructing ontologies and using them as a part of applications, enabling systems also include application programming interfaces (APIs) for writing programs to interact with OWL and RDF (Resource Description Framework) [W3C Recommendation, 2004b] content, such as Jena[3].

EXAMPLE: DEVELOPMENT AND DEPLOYMENT OF AN ONTOLOGY FOR ORAL MEDICINE

In the Swedish Oral Medicine Web (SOMWeb) project, an ontology was developed for representing oral medicine examinations. This work is based on the MedView project [Jontell, et.al, 2005], and its present definitional knowledge representation. This section will describe how the project has utilized the stages and some of the system life cycle processes described above in the ontology development.

1. Concept stage

   Stakeholder Requirements Definition Process: Stakeholders identified were clinicians and IT personnel at the clinic of oral medicine (domain experts, users, and maintainers), and persons from computer science departments (ontology developers and maintainers). A motivating scenario was using the ontology as a schema to represent examinations in oral medicine in RDF for the SOMWeb online community. The purpose of the ontology is to represent concepts relevant to examinations in oral medicine, and the scope of the ontology is to be able to represent at least that which can be represented with MedView's previous knowledge representation. Important terms are foremost those already enumerated by the previous representation; different parts of examinations and properties associated with these, as well as value lists for the properties.

2. Feasibility stage

   Acquisition Process: Unfortunately, there were few ontologies which could accommodate the domain-specific needs and none available in Swedish. Since the use of the W3C standards was decided on early in this work, we searched for medical and dental ontologies which could be considered relevant and which were represented in OWL. No relevant domain-specific OWL ontologies were found, although some had fragments translated into OWL. The ontology identified for reuse, Dublin Core[4] represents meta-data. Automatic

---

1 http://www.w3.org/2001/11/IsaViz/
2 http://jena.sourceforge.net/
3 http://dublincore.org/
4 The general anamnesis is the medical history of the patient.

or semi-automatic ontology construction was not considered outside the use of MedView's knowledge content.

3. Development stage

Acquisition Process: It was decided that translating the larger medical ontologies, not in OWL, into OWL, was beyond the scope of this ontology development project. The ontology identified for reuse, Dublin Core, is small and already in OWL, thus no adaptions were made.

Architectural Design Process: The list compiled from the Stakeholder Requirement Process in the Concept stage was used to decide what should be represented as classes, properties, and instances. It was decided that the different parts of the examination, such as *general anamnesis*□ and *diagnosis*, should be represented as classes. Associated with each of these are properties, such as *hasAllergy* and *hasTentativeDiagnosis*. These properties can take values from instances of corresponding value classes, such as *Allergy* and *Diagnosis*, respectively.

Implementation Process: It was decided that OWL DL should be used, due to being an W3C recommendations, which hopefully will make it easier to integrate SOMWeb's collected data with data from other sources, and give access to a greater range of tools developed in accordance with the recommendations. The sublanguage of OWL DL was chosen as there was a potential need for cardinality constraints more advanced than those offered by OWL Lite, and OWL Full was not an option since computational guarantees are required.

Verification Process: For the development of the SOMWeb ontologies, reasoning test cases were not used, since the logical complexity of the ontologies was not high.

Validation Process: The ontology developers discussed the ontology design with the domain experts. Since much of the knowledge acquisition had been carried out in the MedView project, such validation had already been carried out previously.

4. Production stage

Implementation Process: The identified classes, properties, and instances were encoded in OWL using Protégé and Jena. A prototype was made more or less manually using Protégé. The final ontology was created programmatically using Jena by reading the examination templates in the old MedView format and creating OWL classes, properties, and instances according to the earlier design decisions.

Verification Process: Since the previous MedView format was used as a specification for the SOMWeb ontologies, and these were automatically translated, the SOMWeb ontologies were found to conform to the specification.

Validation Process: In the Protégé prototype we found that the SOMWeb ontology could be used to fulfill the motivating scenario of providing templates for examination records.

Maintenance Process: The importance of end-user development means that it should be possible for the users to add instances to the ontology, but metadata should be added to show who has created a concept and ideally for what purpose. In SOMWeb, the purpose for adding an instance is because it is needed in an examination record. A versioning policy has not yet been established.

5. Utilization stage

Operation Process: The ontology is used as a schema for representing medical examinations in the SOMWeb system [Falkman et al., 2008]. Other parts of this system, such as data about members, meetings, news, and case metadata, are also represented using OWL and RDF.

Supply Process: The ontologies are available at:http://www.somweb/ontologies/ . However, the instance data from examinations cannot be publicly available. The examination instances entered in the online community are available online through member login, but when the clinicians view the cases, they will see natural language representations rather than the RDF.

Maintenance and retirement stages are not included here as the development has not yet reached those stages.

DISCUSSION

Using ISO/IEC 15288 for ontology development bears similarities to other methodologies for ontology construction, such as those described above, being most similar to the Methontology approach. However, the applicability of the standard to a wide range of system related activities brings about a desirable level of adaptability to different ontology development configurations. An ontology will probably not exist in isolation, but will be part of a larger system, composed of both software components and people. Since ISO/IEC 15288 can be applied to all levels of the encompassing system, it makes sense to also apply it at the ontology-level. The standard gives a framework of common system-level thinking and action, regardless of technology and discipline. In viewing ontologies as systems, we acquire access to an interesting and useful conceptual framework, with a holistic view of software and systems engineering. The life cycle process standard provides a basis for stage-based life cycle models and a tailorable process framework, which provide starting points for communication and coordination. A possible drawback of using such a comprehensive standard is that you may have to deal with more factors than are relevant to ontology development. However, the tailorability of the standard should help cope with this problem, so that processes not relevant can be tailored out.

Support is provided by ISO/IEC 15288 for technical systems, non-technical systems, and for systems consisting of both technical and non-technical system elements. This is compelling in the case of ontologies, where non-technical system elements, such as human cognition and agreement on domain conceptualizations, are very relevant. In continuing on the Systems Thinking path, it would be very interesting to consider Senge's [Senge, 1990] concepts on 'dialogue' and 'skillful discussion' for reaching consensus regarding ontological concepts. Dialogue is defined as "sustained collective inquiry into everyday experience and what we take for granted". In skillful discussion – differing from unproductive discussion in that participators do not solely taking part in 'advocacy wars' – a range of techniques, such as collaborative reflection and inquiry skills, are used. Another tool presented in [Senge, 1990] is the 'links and loops'-language, where representations of multiple causal relationships can be constructed using the simple notions of links, loops, and delays. Using this to analyze the problems encountered in developing and deploying ontologies in different situations might prove insightful.

Working within the framework of ISO/IEC 15288 gives access to detailed guidelines for the different stages of development, though these are not specific for ontologies. A comprehensive mapping for ontologies has not yet been completed, but is an interesting work for the future. Also, the proposed processes can be elaborated to include greater detail for the different aspects of ontology management, e.g., elicitation, maintenance, and evolution, based on current research in these areas. A relevant extension of the work presented here would be to outline how the standard and its processes can be applied to the development of ontology-based software.

CASE STUDY REFERENCES

Falkman, G., Gustafsson, M., Jontell, M., and Torgersson, O. (2008) SOMWeb: A Semantic Web-based System for Supporting Collaboration of Distributed Medical Communities of Practice. Journal of Medical Internet Research 10(3):e25, Theme issue on Medicine 2.0.

Fernández-López, M., Gómez-Pérez, A and Rojas Amaya, M. D. (2000) Ontology's crossed life cycles. In EKAW '00: Proc. 12th European Workshop on Knowledge Acquisition, Modeling and Management, pages 65–79, London, UK.

Gruber, T.R. (1993) A translation approach to portable ontologies. Knowledge Acquisition, 5(2):199–220

Grüninger, M. and Fox, M. (1995) Methodology for the design and evaluation of ontologies. In IJCAI'95, Workshop on Basic Ontological Issues in Knowledge Sharing.

D. Gómez-Pérez and Rojas-Amaya, D. (1999) Ontological reengineering for reuse. In EKAW '99: Proc. 11th European Workshop on Knowledge Acquisition, Modeling and Management, pages 139–156, London, UK.

Gustafsson, M. (2006) Ontology Development and Deployment Using ISO/ IEC 15288 System Life Cycle Processes. In Joachim Baumeister and Dietmar Seipel (eds.): Proceedings of the Second Workshop on Knowledge Engineering and Software Engineering, June 14-19, 2006, Bremen, Germany. pp. 15-26.

Gustafsson, M. (2008) SOMWeb: Supporting a Distributed Clinical Community of Practice Using Semantic Web Technologies, PhD Thesis Chalmers University and Göteborgs University.

Lytras, M. D. (2004) Tom gruber in ais sigsemis bulletin! AIS Special Interest Group on Semantic Web and Information Systems, 1(3).

Noy, N.F. and McGuinness, R.L. (2001) Ontology development 101: A guide to creating your first ontology. Stanford Knowledge Systems Laboratory Technical Report.

Pinto, H.S. and Martins, J. P. (2004) Ontologies: How can they be built? Knowledge and Information Systems, 6(4):441–464.

Schreiber, G. and Wood, D. (co-chairs) (2004) Semantic Web Best Practices and Deployment Working Group. W3C Working Group. Available at: http://www.w3.org/2001/sw/BestPractices/

Uschold, M. and King, M. (1995) Towards a methodology for building ontologies. In Proc. IJCAI95's workshop on basic ontological issues in knowledge sharing, Montreal, Canada.

W3C Recommendation (2004a) Web Ontology Language (OWL). See http://www.w3.org/2004/OWL

W3C Recommendation (2004b) Resource Description Framework (RDF). See http://www.w3.org/RDF

Chapter 8
Organizations and
Enterprises as Systems

Achieving Purpose, Goals and Mission

In Chapter 1, a frame of reference for this book was established; namely that in order to achieve purpose, goals, and missions, organizations and their enterprises must have a system focus. This final chapter considers the integration of the sustained system assets of an organization. Remember, as described in the first chapter, the terms organization and enterprises are considered to be equivalent. The integration of the elements of an organization results in an organizational System-of-Interest as an aggregate of systems. Thus, systems are viewed from a business management perspective [Arnold and Lawson, 2004]. They are the building blocks that define a systems landscape and provide substance to the Organization or more popularly called the Enterprise Architecture.

ORGANIZATIONAL SYSTEM(S)-OF-INTEREST

The selection of the systems to become members of the institutionalized system asset portfolio of an organization is based upon what the organization deems their needs to be in terms of operating their business. That is, the organization System-of-Interest is composed of those systems that will collectively deliver the required services. When the organization is in operation based upon instantiation of the defined organization System-of-Interest, it is expected to deliver the desired effect; namely achieving purpose, goals and missions. Thus, the entire organization when in operation is a Respondent System and responds to the Situation System of its "customers" needs by delivering value added products and/or services.

While the actual systems vary amongst organizations, system categories that can be found in one form or another within most all public, private and non-profit organizations are portrayed in Figure 8-1. Organizations are provided with form and content via their infrastructure elements consisting of people, facilities, processes, methods, procedures and data-information-knowledge. Infrastructure elements are utilized as enabling systems in producing value-added products and services as well as in providing organizational business management. Examples of various types of value-added products and services produced by organizations were presented in Chapter 6.

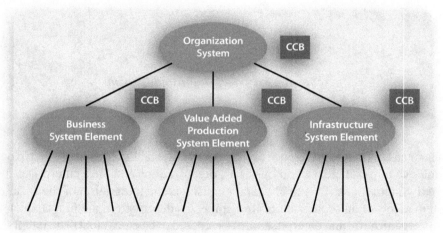

Figure 8-1: Organization System-of-Interest

Each of the system elements of the organization System-of-Interest become, via recursive decomposition, a System-of-Interest itself composed of system elements that meet specific needs, provide necessary services and when operated deliver desired effect. As with all systems the need for further decomposition of the organization system is based upon risk and cost/benefits concerns. That is, that the lowest level systems meet all of the required needs and that the organization management is convinced that there is no undue concern (risk) related at this final level. The systems when operated are expected to deliver their services in a reliable manner.

The hierarchical architecture portrayed in Figure 8-1 illustrates instances of an organization defined Change Management System as indicated in the figure by the CCBs. The CCB's have authority and responsibility for the operational and structural change management decisions that are relevant for their respective areas of the organization.

An organization is often portrayed in the form of an organization chart within a hierarchy of functions such as Sales, Marketing, Advertising, Customer Support, Production, Warehousing, Transport, Human Resources, and so on. However, as an alternative and in striving toward thinking and acting in terms of systems, it is beneficial to view the organization from the perspective of the cooperating and enabling systems that are defined and operated within the organization. Thus, their actual relationships are not necessarily hierarchical, but follow the networking needs for cooperation and support. It is in this interrelationship that thematic systems can be extracted from the institutionalized systems and form the basis for studying both problem and opportunity situations.

Kaleidoscope View

Within an organization there exist many views of the systems that are important for it to operate in a successful manner. Some potential views are as follows:

- the value added products and/or services provided
- the set of agreements it makes and maintains
- the resources (human, financial, and physical)
- the systems and their life cycle models
- the processes that are defined and utilized
- the organizational structure of authorities and responsibilities

All of these views lead to various types of systems which when viewed by individuals and groups define System(s)-of-Interest from a particular view. In a learning organization, it is vital to unify these and other relevant views in order to form a shared vision in order to provide for the cooperation required in achieving a success.

Managers as System Owners

In taking a system's view of an organization, the more traditional definition of management functions is replaced by (or is defined to include) a system owner role as was described in Chapter 4. Ownership can involve system definitions, ownership of production, or ownership of one or more instances of a produced system. Based upon the type of system involved, system ownership can involve a wide variety of management roles as illustrated in Table 8-1. The reader can substitute "Manager of the X System" to identify the human who owns the corresponding system.

Table 8-1: System Management Roles

Asset System	Business System
Change Management System	Configuration System
Contract System	Data System
Engineering System	Facilities System
Financial System	Human Resource System
Information System	Intellectual Property System
Investment System	IT System
Knowledge System	Life Cycle Process System
Logistics System	Marketing System
Policy System	Process System
Product System	Production System
Program System	Proposal System
Public Relations System	Quality System
Requirements System	Resource System
Risk System	Sales System
Security System	Service System
Strategic System	Supply Chain System
Technology System	Waste System

Distribution of Authorities and Responsibilities

The managers in their role as system owners assume responsibility for and should be vested with the authority to life cycle manage their systems. In order to provide effective management the system owner should utilize an instance of a Change Management System in the form of a CCB composed of representatives of those parties who are directly affected by decisions as well as appropriate expert advisors. A hierarchy of CCBs evolves that to a large extent corresponds to the system structure of the organization. Within this structure, authorities and responsibilities are passed to lower levels as portrayed in Figure 8-2.

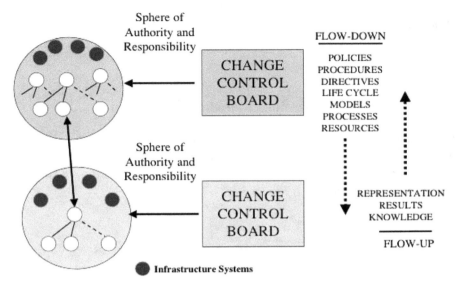

Figure 8-2: Distribution of Authority and Responsibility

In a System-of-Interest that is life cycle managed by a higher level CCB, a system element is supplied by a lower level organizational entity. Thus, the relationship between these two organizational entities is that of an acquirer-supplier that can (and should) be regulated using the guidance from the ISO/IEC 15288 Agreement processes. Note the similarity to the supply chain discussion in Chapter 4. In this case the acquirer and supplier are members of the same organization.

As authority and responsibility for the life cycle management of systems as shown in Figure 8-2 is distributed throughout the organization, various flows arise as a natural consequence of the acquirer-supplier relationship. Some of the more important flows are indicated to the right. In the case of flow-down there are several important aspects that are related to the application of the Organization Project-Enabling processes of ISO/IEC 15288. One should remember the discussion in Chapter 5 concerning the use of the Organization Project-Enabling processes in change management.

As a part of the flow-up, it is important to note the aspect of *representation*. In line with the notions of Russel Ackoff's [Ackoff, 1994] organization (see Chapter 5), it is important that the parties representing lower level interests participate in the decision making of higher levels that will affect their work. Ackoff refers to this as the democratic hierarchy.

Hierarchical relationships of CCBs that evolve follow the System-of-Interest structuring as portrayed in Figure 1-9 and Figure 5-8. That is, there is recursion in the organization, as found in Stafford Beer's Viable System Model [Beer, 1985], that reflects the recursive decomposition of systems along the lines of agreements forming acquirer-supplier relationships (see Chapters 4 and 5).

Implementing an organizational system in the manner described above provides real substance to the system approach to management principle that is called for in the ISO 9001 quality management system standard [ISO 9001].

Organizations in their Environments

The Organization or its Enterprises should be viewed as a Narrow System-of-Interest (NSOI) that exists in a WSOI (Wider System of Interest) and collectively exists in an Environment and even Wider Environment as was introduced in Chapter 1 and now portrayed again in Figure 8-3.

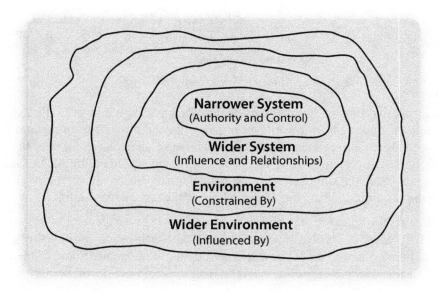

Figure 8-3: Organization as a Narrow System-Of-Interest

As noted in the figure an organization or Enterprise within their sphere have authority and control over the system and its operation. That is, they control their own destiny and make related decisions. They can only influence and have formal or informal relationships with the WSOI that can, as discussed in the toy manufacturer example in Chapter 2, consist of their customers and their supply chain. The combined NSOI and WSOI organization(s) are constrained by factors established in the Environment can also be influenced by factors established by a Wider Environment, for example, rules and regulations, and so on.

ENTERPRISE ARCHITECTURES

"An Enterprise Architecture is a set of descriptive representations (i.e. models) that are relevant for describing an Enterprise such that it can be produced to management's requirements and maintained over the period of its useful life"

John A. Zachman, Zachman Institute for Framework Advancement [www.zifa.com]

In an effort to identify how organizations and their enterprises could exploit the advantages of information technology, Zachman in the 1980s developed a framework model for identifying essential aspects of an organizations information needs as representations in the form models. [Zachman, 1987 and 2008] Thus, his first efforts were referred to as Information Architectures, but later via various refinements the models were collectively called Enterprise Architecture. In reality, Zachman provides a framework for guidance in the relationship between the two, that is, the Enterprise and its Information Architecture.

In Chapter 4 we considered Architecture Frameworks (AF) as described in the ISO/IEC 42010 standard that provides a basis of establishing organizational standards for describing and communicating the essential structures; for example within a group responsible for product line management, project management or an enterprise. The Zachman Framework as well many of the other frameworks such as DoDAF, MoDAF, NAF, FEAF, TOGAF are being utilized in defining the architecture of an Enterprise. Many of the frameworks have been defined by committees are quite large and complex. As mentioned earlier, it is interesting to note that the descriptions of many of these AFs are several hundred pages.

An INCOSE Fellow makes the following pertinent observation:

"I have been monitoring the buzz regarding enterprise architectures, the Federal Enterprise Architecture and the DoD Architecture Framework, DoDAF, as well as numerous commercial enterprise architectures and architecture frameworks. Overall, I have not seen this many lemmings, cliff-side, since data modeling was supposed to save us all."

Jack Ring, INCOSE Fellow and Agent Provocateur

The complexities indicated by Ring have been apparent in utilizing DoDAF and MoDAF. The diversities of viewpoint descriptions have given rise to a multitude of model types for the various views. Thus, as was indicated in Chapter 3 (*"Standards are generally required when excessive diversity creates inefficiencies or impedes effectiveness"*), the Object Management Group has developed a more restrictive set of viewpoints based solely on UML and SysML [OMG, 2009]. The usage of

UPDM (the Unified Profile for the Department of Defense Architecture Framework (DoDAF) and the Ministry of Defence Architecture Framework (MODAF)) is described by [Hause and Holt, 2010].

Applying the Light-Weight Architecture Framework

Given the perspective on systems provided in this book, one can make the observation that a consistent manner of viewing the Enterprise is by considering it to be the aggregate of all of the systems that are interest to the Enterprise as well as the interrelationships amongst the systems. The Enterprise Architecture is then the architectural descriptions of the systems-of-interest. A representative set of such systems was presented in Table 1-1. This is now portrayed in the form of the hierarchy portrayed in Figure 8-4. Perhaps the word Enterprise System is useful in this context to differentiate the enterprise as an operational system in contrast to the description of the system and its elements that constitutes the Enterprise Architecture.

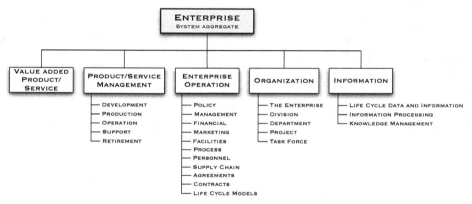

Figure 8-4: Enterprise System as an Aggregate of Systems

Thus, the Enterprise System is expressed in terms of the value added products and or services it provides as well as the systems that it utilizes in the organization in order to produce its product(s) and/or service(s). All of the systems of the Enterprise System portfolio must be managed in such a manner that when they are required to be used in responding to relevant situations, will operate properly. This is consistent with the view presented earlier in this chapter that managers of various functions in an enterprise are in fact system owners.

So, from this systems perspective the LAF (Light-Weight Architectural Framework) introduced in Chapter 4 can be applied in developing the composite architecture of an enterprise. Remember that LAF builds upon the application of the ISO/IEC 15288 and ISO/IEC 42010 standards as well as the concrete system semantics presented in the form of the Systems Survival Kit.

According to LAF, the stakeholders (owner, conceiver, developer, producers, user and maintainer) that have particular interests in the life cycles in respect to (capabilities, requirements, functions/objects, product/service and service utilization and support) express their viewpoints in developing the model views for each system. It is based upon these views that the Enterprise Architecture can be defined.

Disclaimer: Work on the Light-Weight Architecture Framework (LAF) is only beginning. It has been proposed as a way to bring order to architecture frameworks in general including Enterprise Architecture Frameworks.

Your authors concern about the complexity of many of the popular Enterprise Architecture Frameworks is shared with others who are also seeking to simplify the frameworks as well as the methods and tools to support Enterprise Architecture work. One of these efforts is based upon the work of Tim O'Neill that has resulted in the ABACUS product provided by Avolution. [www.avolution.com] This effort started by supporting the IEEE 1471 standard that was the predecessor of the ISO/IEC 42010 standard which is one of the cornerstones of the Light-Weight Architectural Framework. A trip to the Avolution web site is highly recommended.

Challenge: For those who are interested in pursuing the LAF ideas, your author recommends that an effort be made to define the Systems-of-Interest for an Enterprise using the LAF notions of viewpoints and views. This really supports thinking and acting in terms of systems. You might consider using the Avolution product ABACUS as a means of creating and maintaining appropriate descriptions. Further, keep your author informed of any work done in this area. THANKS in advance.

LEADING ORGANIZATIONAL CHANGE

The fundamental elements of Change Management with respect to organizational cybernetics were presented in Chapter 5. The scope of change management includes decisions related to changes in operational parameters as well as the more fundamental structural changes in an organization. In this section, focus is placed upon the leadership of strategic structural change that as noted in Chapter 6 involves taking the actions aimed at improving future operations in respect to solving problems and/or pursuing opportunities. Achieving this requires a well developed process involving the clear identification of the current situation, establishing goals for change, understanding and analyzing the gap between the present and the desired future and then implementing appropriate changes via Respondent System projects.

Certainly such fundamental changes should be made in the context of thinking and acting in terms of systems within a learning organization. Thus, while the skills related to thinking and acting in terms of systems should permeate the organization, they are of the utmost importance for Change Control Boards (CCBs). The individual members of a CCB as well as the CCB collectively should learn to and practice the five disciplines identified by Senge and described in Chapter 7; namely:

- Personal Mastery,
- Mental Models,
- Shared Vision,
- Team Learning and
- Systems Thinking

In practicing these disciplines, they should be able to identify problems and opportunities that are to be addressed by studying thematic systems and exploit languages and methodologies such as those of Senge, Checkland and Boardman as described in Chapter 2. The effective use of Creative Thinking disciplines such as the analogical reasoning of Syntectics, described in Chapter 7, also provide a basis for moving from Knowledge to Wisdom in identifying the potential futures to be obtained by structural change.

In respect to acting in terms of systems, the leadership qualities of the CCB must include the ability to structure change activities, initiate change activities, monitor the change activities, take required corrective actions based upon situations arising during change, and finally see to it that the structural changes are transitioned into the organization in such manner that the desired effect of change is achieved. These vital aspects are treated via the application of the ISO/IEC 15288 standard, as introduced in Chapter 3.

Process of Strategic Changes

As noted in Chapter 5, the ISO/IEC 15288 standard does not provide any separate form of process for Change Management. An approach to implementing a Change Management Process is its creation according to the Tailoring Process of ISO/IEC 15288. In Chapter 5, a proposed purpose for such a process was defined as follows:

Change Management	make decisions related to and provide for the control of changes of any nature that are essential to achieving enterprise purpose, goals and missions

In focusing on the leadership of structural change, the outcomes and activities introduced into the Change Management Process must reflect the fundamental nature of making such strategic changes. Flood [Flood, 1999] presents a process for organizational learning and transformation based upon models of change that can be used in developing a Change Management Process. A version of the Flood's proposed model is presented in Figure 8-5.

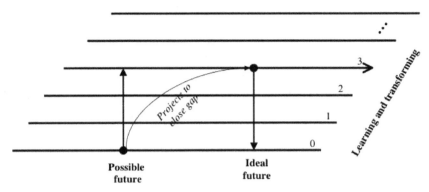

Figure 8-5: Leadership of System Changes

The figure portrays a series of time steps during which possible futures of the organization and/or enterprise evolve. In learning and transforming, a change management entity such as the CCB must continually monitor change triggers by utilizing OODA loop of the Change Model introduced in Chapter 3. The CCB must identify both possible futures and desirable or ideal future situations in the form of a shared vision. The differences between the possible and the ideal futures are then identified through a gap analysis. The CCB establishes and monitors gap-closing Respondent System projects that follow a PDCA loop in making the structural changes in the organization system. As time progresses, the project re-sults, in respect to the life cycles of the systems under change, are measured and evaluated at stage decision gates. In the following temporal period, steps are to be taken to regulate existing and perhaps initiate additional gap-closing projects to achieve an updated ideal future as portrayed in Figure 8-6. In accordance with the Change Model, as this change process takes place knowledge is gathered that is both fed back to the CCB and fed forward to Respondent System projects and line organizations.

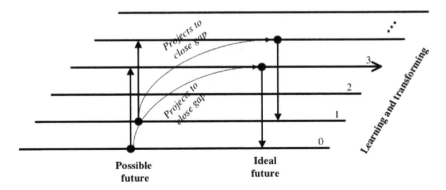

Figure 8-6: Leadership of System Changes over Time Periods

Omnipresence of Cybernetics

Desired ideal futures become the thresholds (benchmarks) against which measured improvement project results are compared in order to revise the project work based upon the current ideal future situation. Thus, the fundamental model of a cybernetic system, as described in Chapter 5, permeates all decision-making of a CCB. In Figure 8-7, the cybernetic structure of CCB related change activities are portrayed.

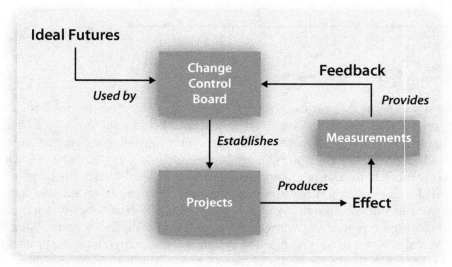

Figure 8-7: Applying Cybernetics to the Leadership of Change

While the cybernetic system model is adapted to the decision-making of CCBs, it is obvious that fundamental decision-making following this model occurs at all individual and group levels in an organization. Cybernetics is an omnipresent phenomenon.

Transforming Organizations

There are many pitfalls associated with making strategic changes in an organization. John Kotter [Kotter, 1990] identifies eight steps to transforming an organization and describes related sources of failures as itemized in Table 8-2.

Table 8-2: Steps and Failures in Transforming an Organization

Steps to Transforming an Organization	*Failures in Transforming an Organization*
1. Establishing a Sense of Urgency – Examining market and competitive realities. Identifying and discussing crises, potential crises, or major opportunities.	Not Establishing a Great Enough Sense of Urgency
2. Forming a Powerful Guiding Coalition – Assembling a group with enough power to lead the change effort. Encouraging the group to work together as a team.	Not Creating a Powerful Enough Coalition
3. Creating a Vision – Creating a vision to help direct the change effort. Developing strategies for achieving that vision.	Lacking a Vision
4. Communicating the Vision – Using every vehicle possible to communicate the new vision and strategies. Teaching new behaviors by the example of the guiding coalition.	Under-Communicating the Vision by a Factor of Ten
5. Empowering Others to Act on the Vision – Getting rid of obstacles to change. Changing systems or structures that seriously undermine the vision.	Not Removing Obstacles to the New Vision
6. Planning for and Creating Short-Term Wins – Planning for visible performance improvements. Creating those improvements. Recognizing and rewarding employees involved in the improvements.	Not Systematically Planning For and Creating Short-Term Wins
7. Consolidating Improvements and Producing Still More Change – Using increased credibility to change systems, structures, and policies that don't fit the vision.	Declaring Victory Too Soon
8. Institutionalizing New Approaches – Articulating the connections between the new behaviors and organization success. Developing the means to ensure leadership development and succession.	Not Anchoring Changes in the Organization's Structure

These steps provide useful guidance for CCBs; in particular, how a CCB interacts with the rest of an organization.

Counteracting the Entropy Effect

In Chapter 5, the entropy effect was described. In managing and leading organizations, it is vital to avoid the increasing entropy effect that deteriorates vital institutionalized system assets. Steps must be taken to provide negative entropy injections in order to assure the availability of stable, reliable, and continually improving systems. The loops and link representation in Figure 8-8 illustrates the injection of negative entropy that can lead to reinforcing loops that reduce entropy as well as a balancing loop that limits the reduction of the entropy effect.

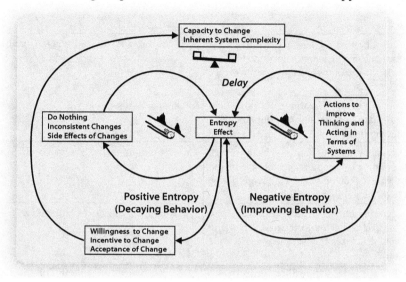

Figure 8-8: Striving after Negative Entropy

The refinement of negative entropy to more fundamental activities and related actions is left as an exercise to the readers. Naturally, this can be based upon the many aspects of thinking and acting in terms of systems that have been introduced during this journey. For example, these include actively utilizing Senge's five disciplines, deploying Checkland's Soft System Methodology, utilizing Boardman and Sauser's Systemigrams, establishing appropriate system architectures, introducing participative democracy in decision-making, applying organizational cybernetics, creating CCBs, implementing ISO/IEC 15288 for system life cycle management, and more.

The sustainability of improvement is also a vital issue. It may be easy to achieve success for so-called *low hanging fruit*; that is, those systems for which change is obvious and where gains can be quickly made. As noted in Table 8-2, Kotter points to advantages of short-term wins, but it is important not to lose sight of the vital aspect of sustaining gains made and striving toward continued leadership and changes in the long run.

ACHIEVING QUALITY IN ORGANIZATIONS, ENTERPRISES AND PROJECTS

An ultimate goal of organizations, enterprises, and their projects is to achieve and retain quality in their value added products and services as well as in their operations. The ISO 9001 standard established eight principles that characterize quality as follows:

Customer focus – Organizations depend upon their customers and therefore should understand current and future customer needs, should meet customer needs and strive to exceed customer expectations.

Leadership – Leaders establish unity of purpose and direction of the organization. They should create and maintain the internal environment in which people become fully involved in meeting the organization's objectives.

Involvement of people – People at all levels are the essence of an organization and their full involvement enables their abilities to be used for the organization's benefit.

Process approach – A desired result is achieved more efficiently when activities and resources are managed as a process.

System approach to management – Identifying, understanding and managing interrelated processes as a system contributes to the organization's effectiveness and efficiency in achieving its objectives.

Continual improvement – Continual improvement of the organization's overall performance should be a permanent objective of the organization.

Factual approach to decision making – Effective decisions are based upon the analysis of data and information.

Mutually beneficial supplier relationships – An organization and its suppliers are interdependent and a mutually beneficial relationship enhances the ability of both to create value.

The reader should easily identify with all of these principles that both directly and indirectly have been considered during this journey through the systems landscape. A detailed explanation of the principles appears on the ISO website http://www.iso.org/iso/qmp.

Implementation of Management System Standards

In addition to Quality Management Systems defined in ISO 9001, a number of other management system standards may be required within an organization. The ISO 14001 Environmental Management System standard is one of them; however there are standards for information security, product safety, supply chain security, occupational health and other areas that also place requirements on management systems. In this section, guidelines for implementing management system standards are provided by demonstrating the approach with respect to the ISO 9000 and ISO 14000 families of standards.

The ISO 9000 and ISO 14000 families of standards are "generic management system standards". That is, the standards can be applied:

- to any organization, large or small, whatever its product or service
- in any sector of activity, and
- whether the organization is a business enterprise, public administration, or a government department.

These two standards have been implemented by over one million organizations in over 175 countries. The requirements of the standards are provided in ISO 9001 [ISO 9001] and ISO 14001 [ISO 14001]. These standards can be utilized for the internal evaluation of an organization and/or as a basis for external including independent certification of conformance to the requirements of the respective standards.

Both ISO 9001 and ISO 14001 promote the utilization of a system approach to the management of quality, respectively of environmental factors. Being systems, they themselves must be life cycle managed by the organization and, hopefully, through some form of Change Control Board. An organization planning to implement a management system according to ISO 9001 and/or ISO 14001 can utilize ISO/IEC 15288 as the basis for managing the life cycles of their management systems. The reflection of this life cycle need and the encapsulation of the system elements of these management systems are portrayed in Figures 8-9 and 8-10.

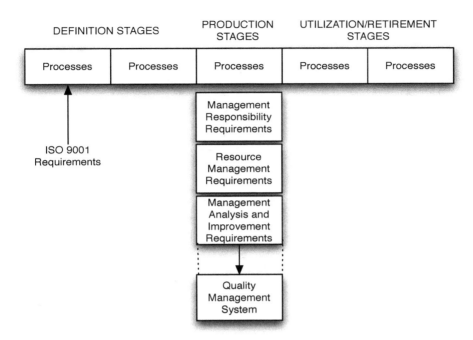

Figure 8-9: Quality Management System Life Cycle and Composition

Note: In the T-diagram as introduced in Chapter 6, the life cycle of a system composed of stages and the processes used in the stages are portrayed horizontally; whereas the system product or service composed of system elements integrated into a system-of-interest is portrayed vertically as a result of some form of "production".

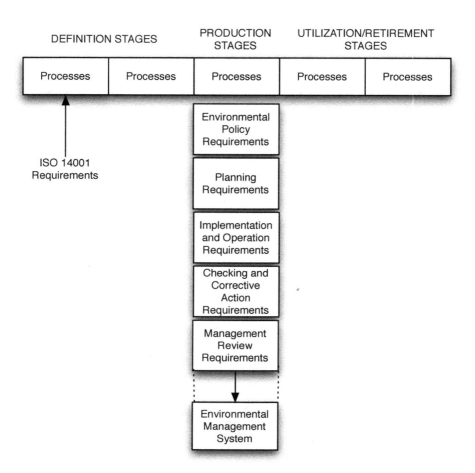

Figure 8-10: Environmental Management System Life Cycle and Composition

The relationships between the system elements are portrayed and defined in the system models provided in the ISO 9001 and ISO 14001 standards, respectively. In both cases, the concrete product of the quality and environmental management systems are documents (i.e. manuals) that are utilized in placing requirements on organization/enterprise activities, products and services. The life cycle management of these management systems provides a structured means of customizing (tailoring) and integrating the requirements of the standards with the realities of an organization. A vital part of this tailoring is the development of policies and procedures. The Enterprise then must assure, via policy and procedures, that system life cycles stages and their constituent processes incorporate, at appropriate places, the requirements of the Quality and Environment Management Systems via the utilization of their respective manuals.

Deploying Quality and/or Environment Management Systems

The deployment (Utilization) of these two management systems involves their use as stakeholder requirement inputs to all other systems that the organization produces in the form of products and/or services or uses as institutionalized system assets for their own business management and infrastructure needs. This utilization is portrayed in Figure 8-11.

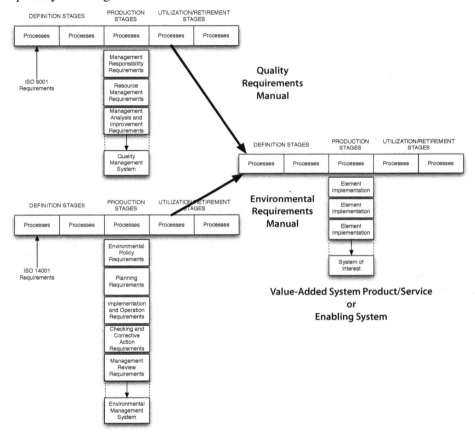

Figure 8-11: Applying Quality and Environment System Requirements

The quality and environmental aspects are taken into account in planning for systems related work that results in successive versions of a System-of-Interest. In particular, in the early stages of the life cycle the requirements are analyzed and then distributed to appropriate stages, processes and activities. Furthermore, the requirements related to product and service properties are provided to Respondent System projects that will perform the actual transformations on system definitions, produce the products and services, operate (perhaps via a line organization) and then potentially dispose of the system product or service. The distribution of requirements is portrayed in Figure 8-12.

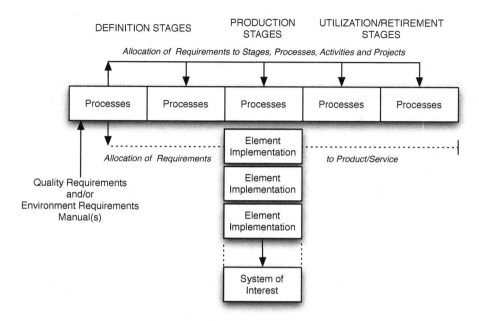

Figure 8-12: Allocating Requirements to Activities, Products and Services

The means of verifying that the requirements of the organization's Quality Management System and/or Environmental Management System have been met for any particular system product or service can be conveniently accomplished by incorporating Verification processes in each relevant stage as portrayed in Figure 8-13. Furthermore, in order to validate that the produced product and/or service actually satisfies the customer, a Validation process is also included in the Utilization stage. Note that the use of Verification processes in the later stages applies to the verification of maintenance or retirement requirements.

DEFINITION STAGES		PRODUCTION STAGES	UTILIZATION/RETIREMENT STAGES	
Processes	Processes	Processes	Processes	Processes
Verification	Verification	Verification	Verification Validation	Verification

Figure 8-13: Verification and Validation

Through the distribution of requirements into life cycles, their verification and validation for customer satisfaction, the framework of ISO/IEC 15288 provides an expedient mechanism for the implementation of ISO 9001 and ISO 14001 as well as other management system standards. Continual management review of the achievement of requirements by CCBs is provided in stage decision gates. Further, when the stages of the life cycle are iterated in the refinement of a system product or service or in producing new versions, the framework continues to provide decision points so that quality and/or environmental requirements remain in focus.

Supporting Project Needs

The requirements placed upon organizations to treat quality, respectively environmental aspects are often transformed into project specific requirements for projects within an enterprise and between enterprises. In such cases, it is important that the project be endowed with the manuals required for their planning, execution of project plans, assessment of results as well as control and revision of project plans. The ISO 10006 standard provides such guidelines for quality management systems in projects [ISO 10006]. Here once again, the utilization of ISO/IEC 15288 provides a framework for treating quality and/or environment requirements within a project. This is accomplished in accordance with the systems approach portrayed in Figure 8-9. The Project Quality Management System is viewed as an enabling system. Thus, in addition to the ISO 9001 requirements (and perhaps ISO 14001 requirements) being propagated additional project specific requirements based on ISO 10006 are used in the early stages of this enabling system.

It is possible that a generic Project Quality Management System is generated and life cycle managed as a Project Quality Manual. In this case, concrete Respondent System projects will use this (and the organization manual) as requirement inputs for each project. Alternatively, a unique project manual can be generated for each project without a generic manual. The course of action depends upon the quantity and diversity of projects that are to be executed. Complex, long term projects, perhaps based upon the involvement of multiple organizations, will most likely require some form of the Project Quality Management System; whereas short term projects may simply adapt necessary aspects from a generic project quality manual and/or the organization quality manual.

KNOWLEDGE VERIFICATION

1. Identify systems corresponding to the system elements of business management, value-added product and/or services and infrastructure in an organization with which you are familiar.

2. How are the systems identified in (1) managed in the organization?

3. What benefits would accrue in the management and leadership of systems identified in (1 and 2) if a Change Control Board was implemented?

4. Identify some additional management roles (see Table 8-1) in which the managers can be viewed as system owners.

5. Describe how strategic planning leadership is accomplished in an organization with which you are familiar.

6. Identify several systems that should be defined in an Architectural Framework for an Enterprise with which you are familiar. Identify the type of models that the stakeholders having interests in each of the systems at various points in their life cycles should utilize.

7. Why is cybernetics and organizational cybernetics so important?

8. Provide some examples where steps have been taken in transforming an organization with which you are familiar. Further describe any failures that may have occurred.

9. Using some examples of thinking and acting in terms of systems, describe how their injection into an organization leads to counteracting the entropy effect.

10. What is low-hanging fruit in terms of leading organizational change?

11. Using the eight quality principles established by ISO 9001, identify relevant knowledge gained during the journey that has addressed these principles.

12. What is the most concrete product of a management system and how is it applied?

13. Why are verification and validation of stage results important?

SUMMING IT ALL UP!!!

"It is in the nature of systemic thinking to yield many different views of the same thing and the same view of many different things."
Rusell L. Ackoff

Now that the journey is coming to an end the reader can appreciate the kaleidoscopic view of systems which corresponds directly with the above quote from one of the pioneers of systems (systemic) thinking.

The journey that has been taken through the various aspects of the systems landscape has provided a well-rounded and holistic view both of thinking as well as acting in terms of systems. As with all journeys, there are several sites that could or should have been visited along the way, but this leaves space for further investigation into this fascinating topic. Your author strongly recommends that you utilize the vast information resources of the Internet to continually gain deeper insight and follow significant new developments concerning systems.

The true test of the usefulness of this journey can be measured in the knowledge that has been attained; the changes that occur as a result of the knowledge, how the knowledge can be applied in the real world, and how the knowledge contributes to personal ethical activities. [Duffy, 1995] provides very useful questions in this regard which you can now use to see what effect this material has had upon your own journey in the systems landscape:

LEARNING AS A "BODY" OF KNOWLEDGE

1. Have I grasped the concepts?
2. How much of the material have I retained?
3. How do you attach the new learning to materials already learned or known?
4. How much more do I now know?

LEARNING AS A CHANGE PROCESS

1. How has my life or behavior changed?
2. Does it change the way I look at the world?
3. Does it challenge my existing knowledge?
4. Has my perception of the course changed since taking the course?
5. What is different from what I knew before? What have I added to what I knew before? What have I revised from what I knew before?

LEARNING AS APPLICATION AND INVOLVEMENT IN THE REAL WORLD

1. Can I use the information in future situations?

2. Will I be able to apply the information outside? In the workplace?

3. Does what I have learned make sense in my life? In my personal life?

4. Am I able to apply the theories in practice?

5. What is the practical application of this course?

6. Can I discuss clearly other examples related to the subject matter discussed?

7. Can I ask a question that would amplify what has just been discussed?

8. Can I explain the idea to someone else?

LEARNING AS ETHICAL ACTIVITY

1. Can I use the information to help others?

2. What information did I deem pertinent enough to integrate into my existing set of beliefs/morals/values?

3. What information did I value enough to pass on?

The journey has come to an end.

Happy travels in the future!!!

REFERENCES

Ackoff, R. L. (1971) Towards a System of Systems Concepts. Management Science, 17(11).

Ackoff, R. L. (1994) The Democratic Organization, Oxford University Press, New York.

ANSI/IEEE 1471-2000 Recommended Practice for Architectural Description of Software-Intensive Systems.

Arnold, S., and Lawson, H. (2004) Viewing Systems from a Business Management Perspective, Systems Engineering, The Journal of The International Council on Systems Engineering, Vol. 7, No. 3, pp 229-242.

ASD (2009) Aerospace and Defence Industries of Europe, ASD-STAN, Products and Services S3000L, www.asd-stn.org.

Ashby, W.R. (1964) An Introduction to Cybernetics, Chapman and Hall, 9, London.

Ashby, W.R. (1973) Some Peculiarities of Complex Systems, Cybernetic Medicine, 9, pp. 1-7.

Beer, S. (1985) Diagnosing the System for Organisations, Wiley, Chichester and New York.

Bellinger, G. (2004) www.systems-thinking.org

Bendz, J. and Lawson, H. (2001) A Model for Deploying Life-Cycle Process Standards in the Change Management of Complex Systems, Systems Engineering, The Journal of The International Council on Systems Engineering, Vol. 4, No. 2, pp 107-117.

Blanchard, B.S. (2004) Logistic Engineering & Management, Megregor, London.

Boardman, J. and Sauser, B. (2008) Systems Thinking – Coping with 21st Century Problems, CRC Press, Boca Raton, FL.

Boardman, J., Wilson, M. and Fairbairn, A. (2005) Addressing the System of Systems Challenge, Proceedings of the INCOSE International Conference, Rochester, NY.

Boehm, B. W. (1988) A Spiral Model of Software Development and Enhancement. IEEE Computer, May.

Box, G. E. P. and Draper, N. R. (1987) Empirical Model-Building and Response Surfaces. Wiley. pp. p. 424

Boyd, J. R. (1987) An Organic Design for Command and Control, A Discourse on Winning and Losing, Unpublished lecture notes (Maxwell AFB, Ala. Air University).

Brehmer, B. (2005) The Dynamic OODA Loop: Amalgamating Boyd's Loop and the Cybernetic Approach to Command and Control: Assessment, Tools and Metrics, Proceedings of the 10th International Command and Control Research and Technology Symposium: The Future of C2. McLean, VA, June 13-16.

Checkland, P. (1993) Systems Thinking, Systems Practice, John Wiley, Chichester, UK.

Checkland, P. (1999) Systems Thinking, Systems Practice – Includes a 30 year Retrospective, JohnWiley, Chichester, UK.

Checkland, P. and Sholes, J. (1990) Soft System Methodology in Action, Wiley, New York.

Checkland, P. and Poulter, J. (2006) Learning for Action, Wiley, New York.

Churchman, (1971) The Design of Inquiring Systems, Basic Books, New York.

Dahl, O-J., Myhrhaug, B. and Nygaard, K. (1970) Common Base Language, Norwegian Computing Center.

DoD (2004) Acquisition Deskbook https://dap.dau.mil

DoD (2005) Technology Readiness Assessment (TRA) Deskbook

Donate, I. (2009) A Systemic View of Technology Readiness Levels (TRL), Student project.

Duffy, M. Sensemaking: A Collaborative Inquiry Approach to "Doing" Learning, The Qualitative Report, Volume 2, Number 2, October, 1995 (http://www.nova.edu/ssss/QR/QR2-2/duffy.html)

Ericsson, M. (2006) Why Customer Product Information is Inadequate, Late and Expensive to Produce or how to make it adequate, ready in time and on budget -a Systemic Approach. Student project.

Fairbairn, A., and Farncombe, A. (2001) Enterprise Systemics: Systems Thinking for Plotting Strategy at the Extended Enterprise Level, INCOSE International Conference, Melbourne.

Flood, R.L. and Carson, E.R. (1998) Dealing with Complexity: An Introduction to the Theory and Application of Systems Science, Second Edition, Penum Press, London and New York.

Flood, R.L. (1998) Rethinking the Fifth Discipline: Learning within the unknowable, Routledge, London and New York.

Fornell, C. (2001) The Science of Satisfaction, Harvard Business Review, 79, 3, March 120-121.

Forrester, J.W. (1975) Collected Papers of Jay W. Forrester, Pegasus Communications.

Forsberg, K., Mooz, H., Cotterman, H. (2005) Visualizing Project Management, 3rd edition, John Wiley and Sons, New York, NY.

Fritzson, P. (2004) Object-Oriented Modeling and Simulation with Modelica 2.1, IEEE Press and Wiley-Interscience.

Gordon, W.J.J., (1961) Synectics, the Development of Creative Capacity, Harper & Row, New York.

Haines, S.G. (1998) The Managers Pocket Guide to Systems Thinking & Learning, HRD Press, Amherst, Mass.

Hammond, E.W. and Cimino, J.J. (2001) Standards in Medical Informatics, Springer

Hause, M. and Holt, J., (2010) Model-Based System of Systems Engineering with UPDM, INCOSE EuSEC Symposium Proceedings.

Herald, T., Berkemeyer, W. and Lawson, H. (2004) Proceedings of the INCOSE Conference, Toulouse, France.

Howard, R. (1960) Dynamic Programming and Markov Processes, The M.I.T. Press.

Howard, R. and Matheson, J.E. (editors) (1984) Readings on the Principles and Applications of Decision Analysis, 2 volumes. Menlo Park CA: Strategic Decisions Group.

INCOSE (2007) Systems Engineering Handbook: A Guide for System Life Cycle Processes and Activities, Version 3.1. See www.incose.org.

Ingargio (2005) http://www.cis.temple.edu/~ingargio/cis587/readings/tms.html

ISO 9001:2008 (2008) Quality Management Systems, International Standardization Organization, 1, rue de Varembe, CH-1211 Geneve 20, Switzerland.

ISO 10006:2003 (2003) Quality management systems - Guidelines for quality management in projects, International Standardization Organization, 1, rue de Varembe, CH-1211 Geneve 20, Switzerland.

ISO 14001:2004 (2004) Environmental Management Systems -- Requirements with Guidance for Use, International Standardization Organization, 1, rue de Varembe, CH-1211 Geneve 20, Switzerland.

ISO/IEC 12207 (1995) Information technology - Software life cycle processes, International Standardisation Organisation/International Electrotechnical Commission, 1, rue de Varembe, CH-1211 Geneve 20, Switzerland.

ISO/IEC 15288 (2002) Information technology – System life cycle processes, International Standardization Organization/International Electrotechnical Commission, 1, rue de Varembe, CH-1211 Geneve 20, Switzerland.

ISO/IEC 15288 (2008) Systems and software engineering - System life cycle processes, International Standardization Organization/International Electrotechnical Commission, 1, rue de Varembe, CH-1211 Geneve 20, Switzerland.

ISO/IEC/IEEE 15939, Measurement Process, International Standardization Organization/International Electrotechnical Commission, 1, rue de Varembe, CH-1211 Geneve 20, Switzerland.

ISO/IEC 15504 (2004) Information technology Process assessment (six parts), International Standardization Organization/International Electrotechnical Commission, ISO,1, rue de Varembe, CH-1211 Geneve 20, Switzerland.

ISO/IEC 19501 (2005) Information technology - Open Distributed Processing - Unified Modeling Language (UML) Version 1.4.2, International Standardization Organization/International Electrotechnical Commission, 1, rue de Varembe, CH-1211 Geneve 20, Switzerland

ISO/IEC 24748-1 (2009) Systems and software engineering - Guide for Life Cycle

ISO/IEC 42010 (2010) Architecture description – Committee Draft 1 (CD1) of the on-going revision.

Management, International Standardization Organization/International Electrotechnical Commission, 1, rue de Varembe, CH-1211 Geneve 20, Switzerland.

Jackson, D, (2009) A Direct Path to Dependable Software, Communications of the ACM, Vol. 52, No. 04, April pp. 78-88.

Jennerholm, M. and Stern, P. (2006) Societal Security Handled by a Crisis Management System of Systems, Student Project.

Joint Publication 2-0 (2007) Joint Intelligence, Joint Chiefs of Staff.

Kaplan, R.S. Norton, D.P. (1996) The Balanced Scorecard: Translating Strategy into Action, Harvard Business School Press, Boston, MA.

Kotter, J.P. (1990) What Leaders Really Do, Harvard Business Review, May-June.

Kotter, J.P. (1995) Leading Change: Why Transformation Efforts Fail, Harvard Business Review, March-April, pp. 59-67.

Kruchten, P. (2003) The Rational Unified Process: An Introduction (3rd edition), Addison Wesley.

Langefors, B. (1973) Theoretical Analysis of Information Systems, Studentlitteratur, Lund, Sweden.

Lawson, H. W., and Martin, J. N. (2008) On the Use of Concepts and Principles for Improving Systems Engineering Practice, INCOSE, Proceedings of the INCOSE International Conference, Utrecht.

Low, A., (1976) Zen and Creative Management, Playboy Paperbacks, New York. ISBN 0-867-21083-4.

Lumina (2009) Influence Diagrams, http://www.lumina.com/software/influence-diagrams.html

Maeir, M., Emory, D., Hilliard, R. (2004) ANSI/IEEE 1471 and Systems Engineering, Systems Engineering, The Journal of The International Council on Systems Engineering, Vol. 7, No. 3.

Markowitz, H.M. (1979) Belzer, Jack; Holzman, Albert G.; Kent, Allen. eds. SIMSCRIPT, Encyclopedia of Computer Science and Technology. 13. New York and Basel: Marcel Dekker. pp. 516.

Martin, J. N. (2000) Systems Engineering Guidebook: A Process for Developing Systems and Products, CRC Press.

MODAF 1.2. (2008) The MOD Architecture Framework, version 1.2 Available from http://www.modaf.org.uk/file_download/39/20081001

Novak, J.D. and Cañas, A.J. (2008) The Theory Underlying Concept Maps and How to Construct and Use Them. Technical Report IHMC CmapTools 2006-01 Rev 01-2008, Florida Institute for Human and Machine Cognition (IHMC) www.ihmc.us

Ntuen, C.A., Munya1, P., Trevino, M., Leedom, D. and Schmeisser, E. (2005) An Approach to Collaborative Sensemaking Process. Proceedings of the 10th International Command and Control Research and Technology Symposium – The Future of C2

OMG, (2009) Object Management Group - Unified Profile for the Department of Defense Architecture Framework (DoDAF) and the Ministry of Defence Architecture Framework (MODAF), available at http://www.omg.org/spec/UPDM/

Pirsig, R.M. (1974) Zen and the Art of Motorcycle Maintenance, William Morrow and Company edition and Bantam Books edition, ISBN 0-553-27747-2.

PMI (Project Management Institute) (2008) A guide to the Project Management Body of Knowledge PMBOK® Guide, Fourth Edition. PMI Inc., Newtown Square, PA.

Rechtin, E. and Maier, M. (2000) The Art of Systems Architecting, CRC Press, Boca Raton, FL. Second Edition.

Rhodes, D.H. and Ross, A.M. (2010) Five Aspects of Engineering Complex Systems: Emerging Constructs and Methods, SysCon2010 - IEEE International Systems Conference San Diego, CA.

Rifkin, J. (1980) Entropy: A New World View, Bantam Books, New York, ISBN 0-553-20215-4.

Roedler, G., Rhodes, D., Jones, C., Schimmoller, H. (2010) Systems Engineering Leading Indicators Guide, Version 2.0, INCOSE-TP-2005-001-03. International Council on Systems Engineering (INCOSE), San Diego, CA.

Rouse, W.B. (2005) Enterprises as Systems: Essential Challenges and Approaches to Transformation, Systems Engineering, The Journal of The International Council on Systems Engineering, Vol. 8, No. 2, pp 138-149.

Royce, W.W. (1970) Managing the Development of Large Software Systems: Concepts and Techniques, Technical Papers of Western Electronic Show and Convention (WesCon) August 25-28, 1970, Los Angeles.

Schreiber, G. and Wood, D. (co-chairs) (2004) Semantic Web Best Practices and Deployment Working Group. W3C Working Group. Available at: http://www.w3.org/2001/sw/BestPractices/

Senge, P.M. (1990) The Fifth Discipline: The Art & Practice of The Learning Organization, Currency Doubleday, New York.

Senge, P.M., Klieiner, A., Roberts, C., Ross, R.B., and Smith, B.J. (1994) The Fifth Discipline Fieldbook: Strategies and Tools for Building a Learning Organization, Currency Doubleday, New York.

Skyttner, L. (2001) General Systems Theory: Ideas and Applications, World Scientific Publishing Co., Singapore.

Stevens, R., Brook, P., Jackson, K., Arnold, S. (1998) Systems Engineering: Coping with Complexity, Prentice-Hall Europe.

Schumann, R. (1994) Developing an Architecture that No One Owns: The US Approach to System Architecture, Proceedings First World Congress Appl. Transport Telematics Intelligent Vehicle-Highway Systems, Paris, France.

von Bertalanffy, L. (1968). General system theory: foundations, development, applications (Rev. ed.). New York: Braziller.

Wasson, C.S. (2006). System Analysis, Design and Development: Concepts, Principles, and Practices, John Wiley & Sons, Inc., Hoboken, New Jersey.

Weaver, W. (1948) Science and Complexity. American Science, 36 pp 536-544.

Weick, K. (1995). Sensemaking in organizations. Thousand Oaks, Calif.: SAGE.

Weinberg, Gerald M. (2001) An Introduction to General Systems Thinking, Dorset House Publishing Company, 1st edition.

Wik, M. (2003) Multisensor Data Fusion in Network-Based Defence, First International Conference on Military Technology, Stockholm,

Wurman, R.S. (1989) Information Anxiety, Doubleday, New York.

Zachman, J.A. (1987) A Framework for Information Systems Architecture, IBM Systems Journal, 26(3).

Zachman, J. A. (2008) The Zachman Framework™: A Concise Definition, Zachman International. Available from http://www.zachmaninternational.com/index.php/the-zachman-framework/26-articles/13-the-zachman-framework-a-concise-definition ver. 1.1.

ABOUT THE AUTHOR

Harold W. "Bud" Lawson has been active in the computing and systems arena since 1958 and has broad international experience in private and public organizations as well as academic environments. Experienced in many facets of computing and computer-based systems, including systems and software engineering, computer architecture, real-time, programming languages and compilers, operating systems, life-cycle process standards, various application domains as well as computer and systems related education and training.

Received the Bachelor of Science degree from Temple University, Philadelphia, Pennsylvania and the PhD degree from the Royal Technical University, Stockholm, Sweden. Contributed to several pioneering efforts in hardware and software technologies at Univac, IBM, Standard Computer Corporation, and Datasaab. Permanent and visiting professorial appointments at several universities including Polytechnic Institute of Brooklyn, University of California, Irvine, Universidad Politecnica de Barcelona, Linköpings University, Royal Technical University, University of Malaya and Keio University. Currently, Honorary Professor in the Swedish Graduate School of Computer Science and Academic Fellow in the School of Systems and Enterprises at Stevens Institute of Technology, Hoboken, NJ.

Fellow of the Association for Computing Machinery, Fellow and Life Member of the IEEE, Fellow of the International Council on Systems Engineering, ACM Distinguished Lecturer, IEEE European Distinguished Visitor, Member of the ACM Fellows Committee (1997-2001), Founding member of SIGMICRO, EUROMICRO, the IEEE Computer Society Technical Committee on the Engineering of Computer-Based Systems, the Swedish National Association for Real-Time (SNART), the Swedish chapters of ENCRESS and INCOSE. Chairman (1999-2000) Technical Committee on the Engineering of Computer-Based Systems. Head of the Swedish Delegation to ISO/IEC JTC1 SC7 WG7 (1996-2004) and elected architect of the ISO/IEC 15288 standard.

In 2000, he received the prestigious IEEE Computer Pioneer Charles Babbage medal award for his 1964-65 invention of the pointer variable concept for programming languages.

Harold Lawson is an independent consultant operating his own company Lawson Konsult AB and is, as well, a consulting partner of Syntell AB, Stockholm.

www.ingramcontent.com/pod-product-compliance
Lightning Source LLC
LaVergne TN
LVHW012328060326
832902LV00011B/1765